Earths Grow Thick

Earths Grow Thick

Including the works *Keys and Cues, Untitled (Gun), How Dickinson Stayed Home,* and *When Dickinson Shut Her Eyes* by Roni Horn. With texts by *bell hooks, Felix Gonzalez-Torres, Judith Hoos Fox, Roni Horn, Amada Cruz,* and *Sarah J. Rogers.*

WEXNER CENTER FOR THE ARTS
The Ohio State University, Columbus, Ohio

Earths Grow Thick

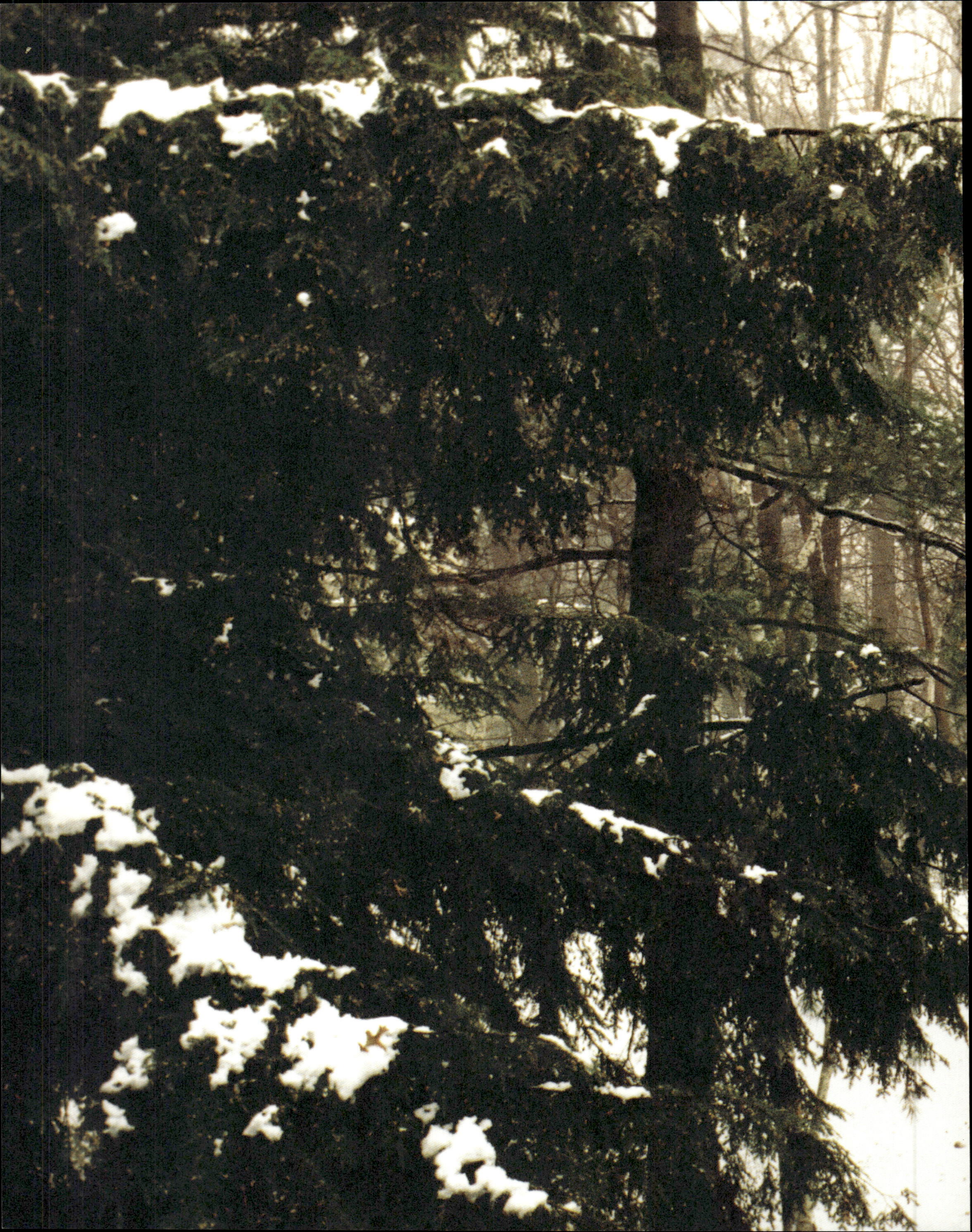

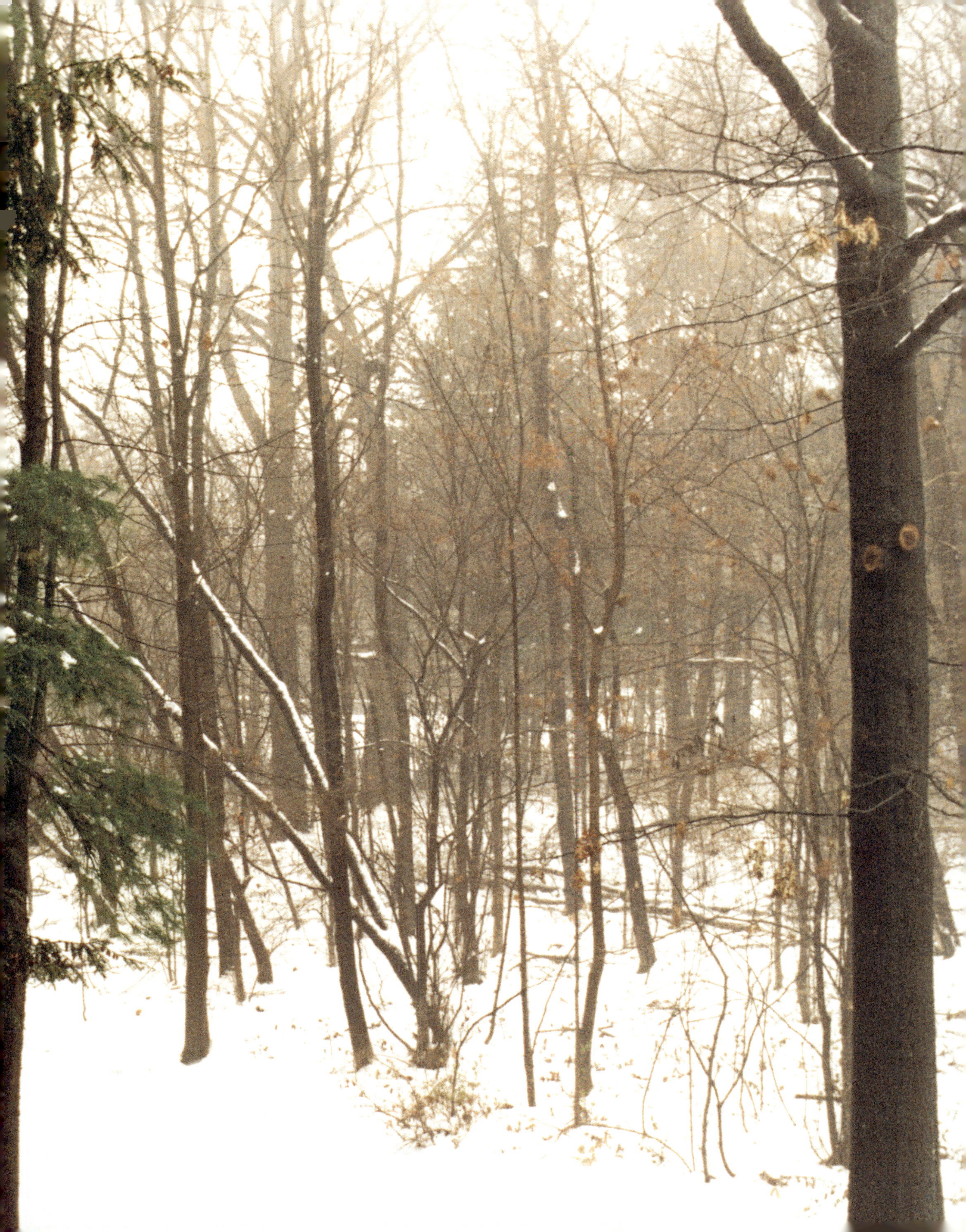

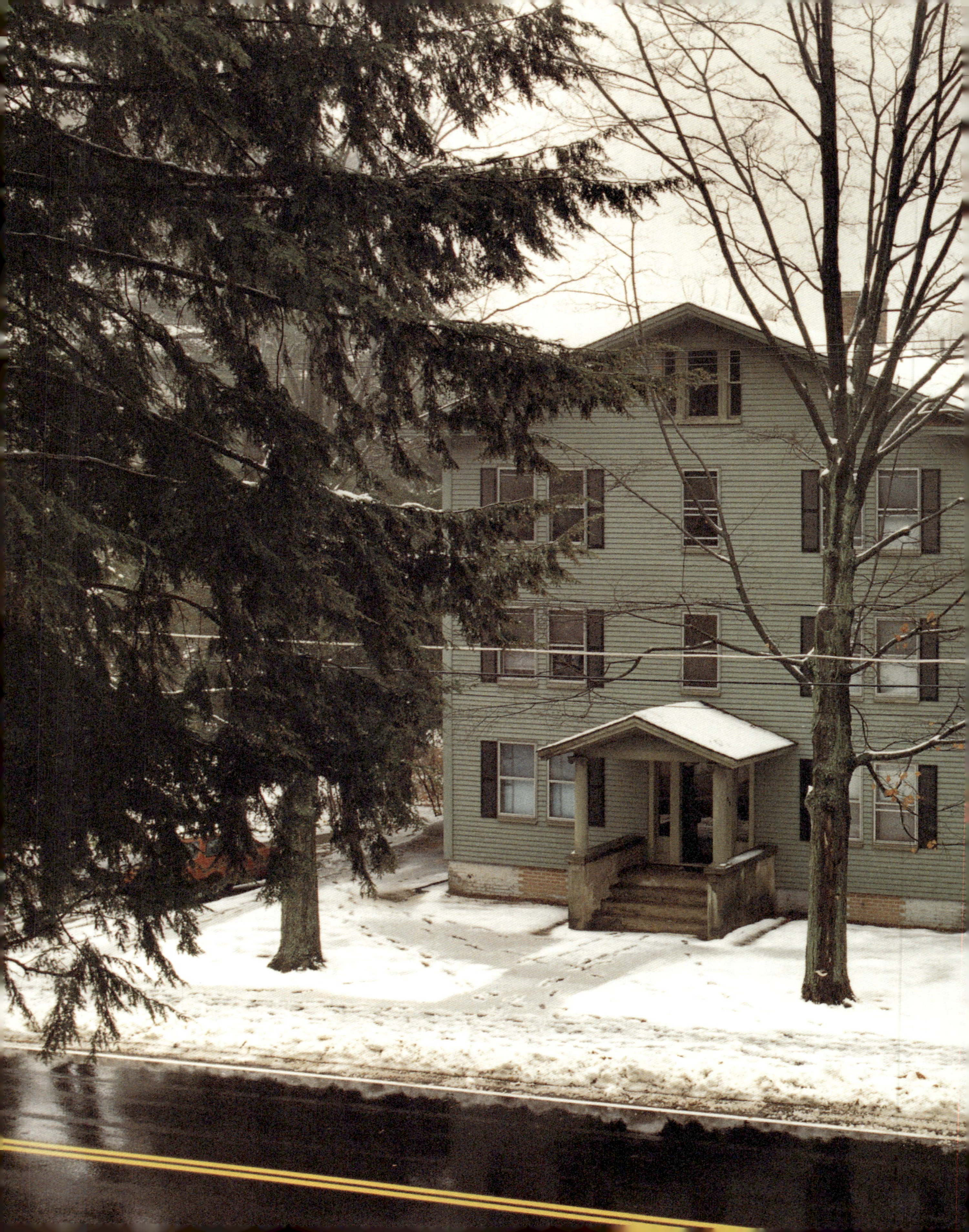

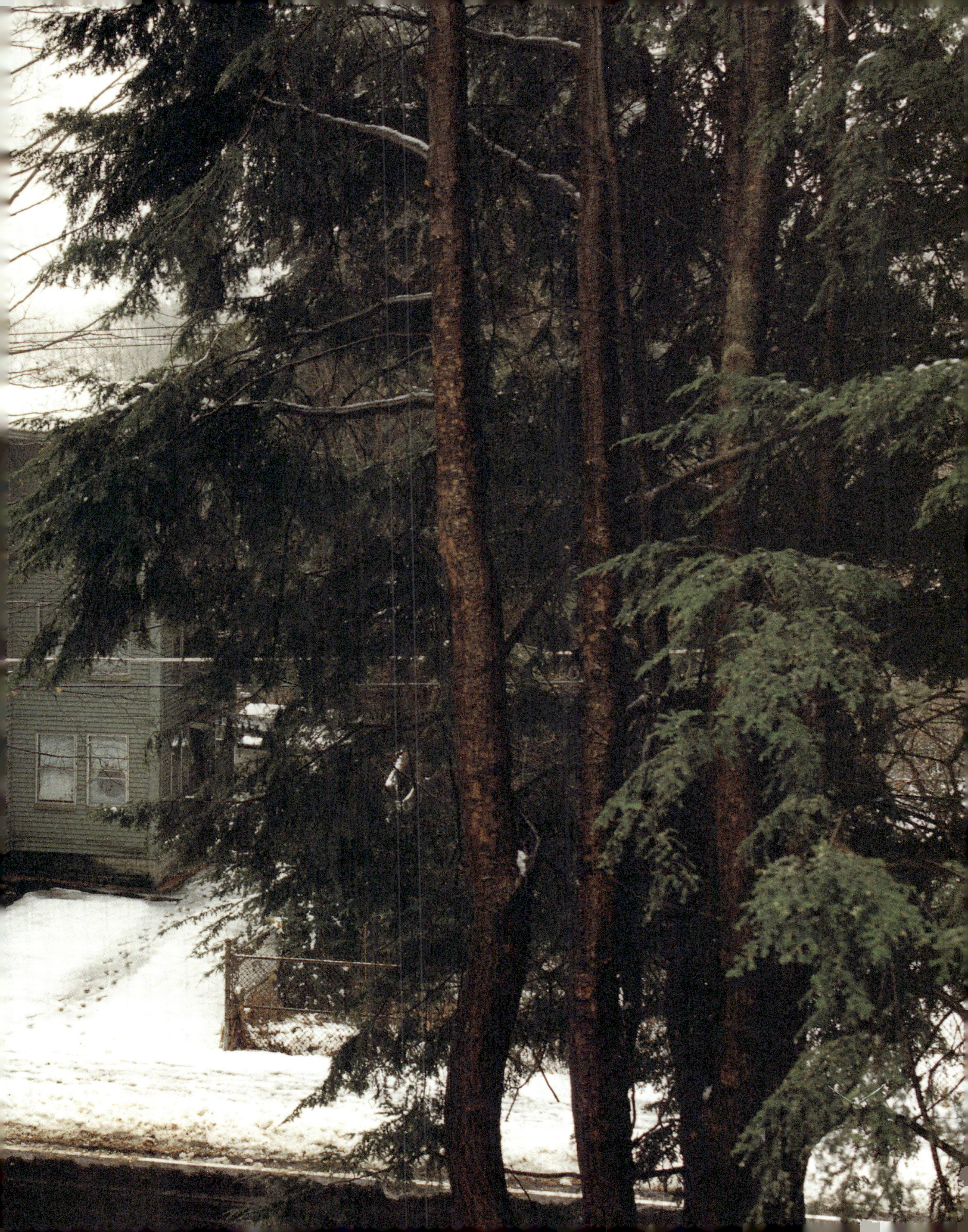

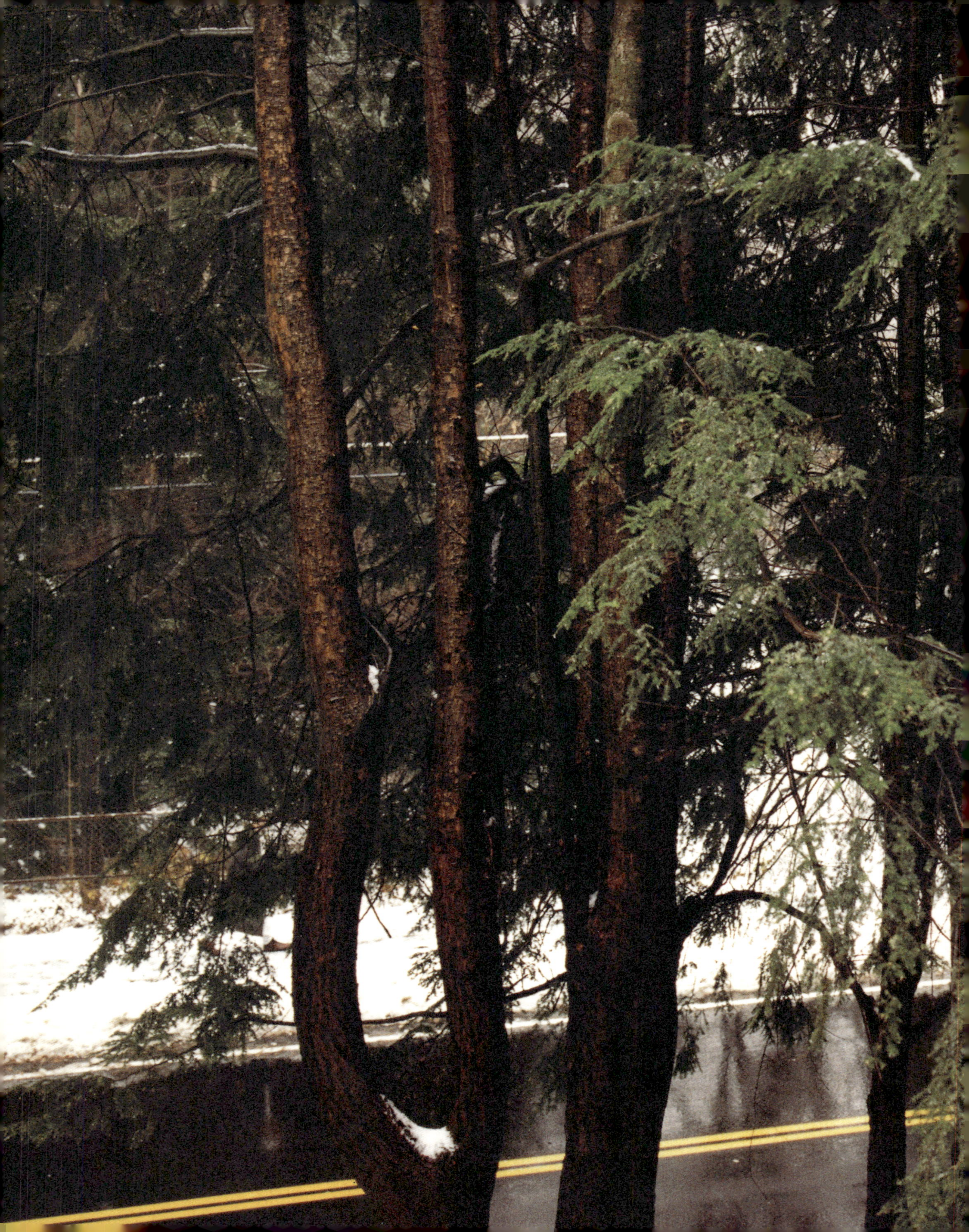

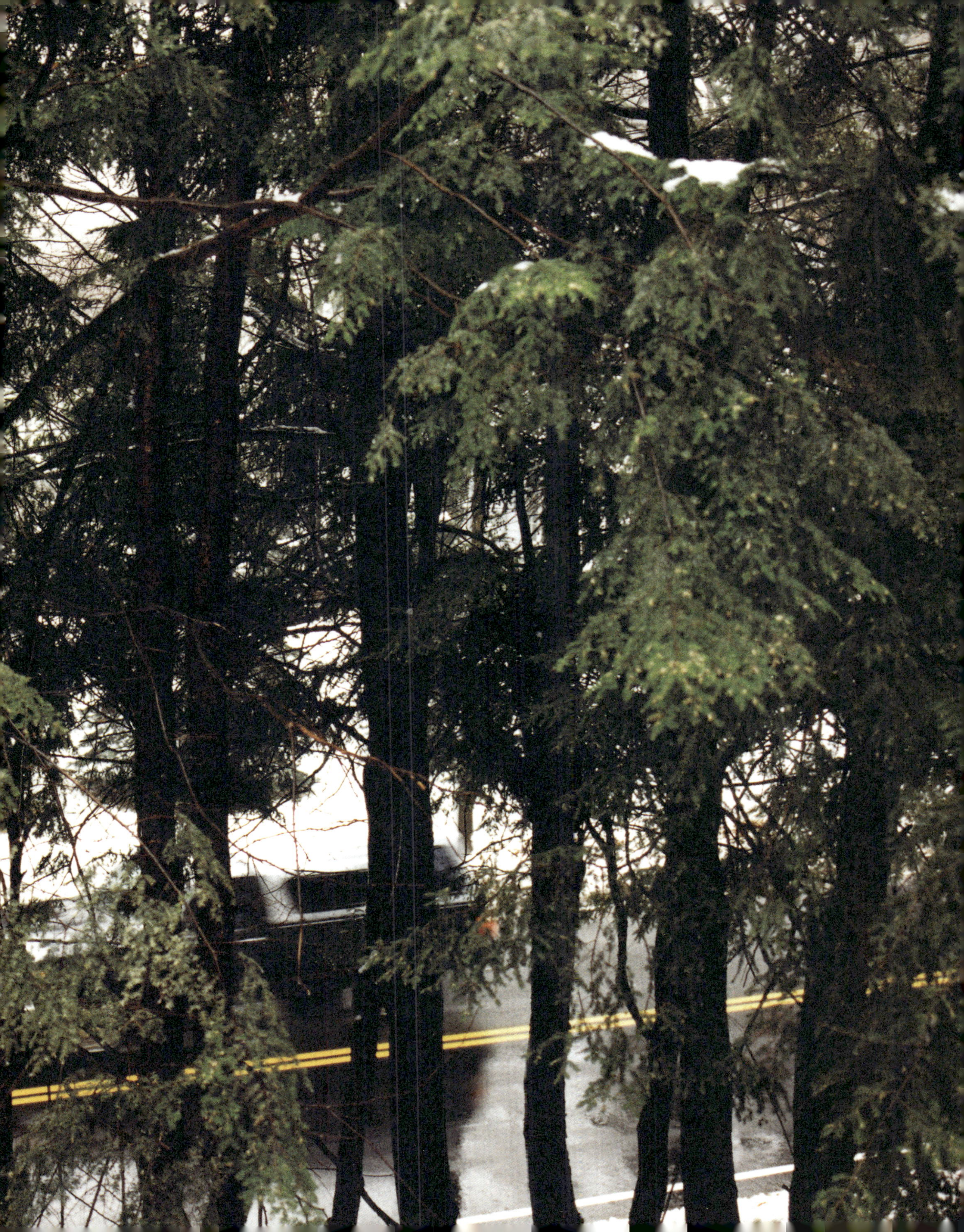

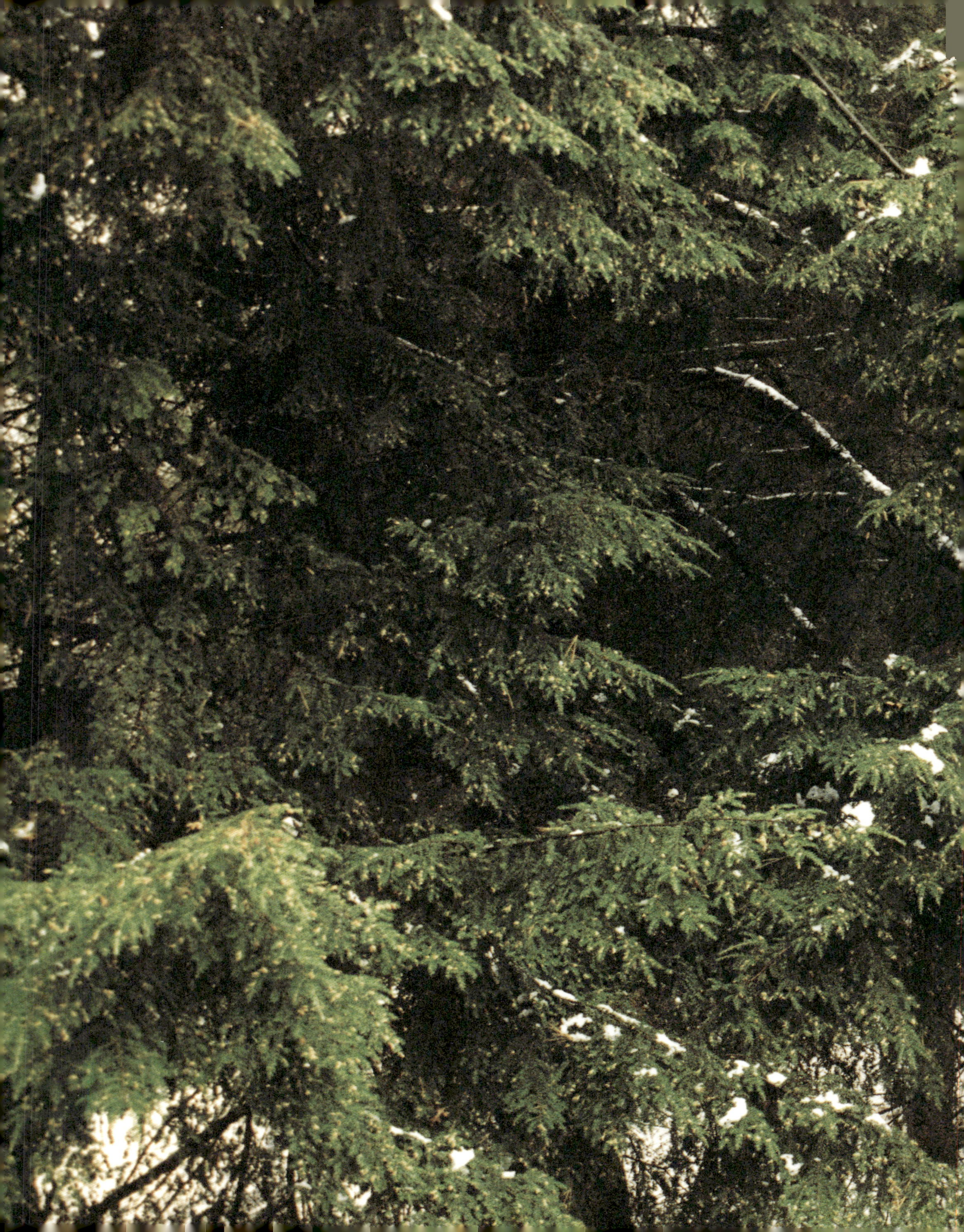

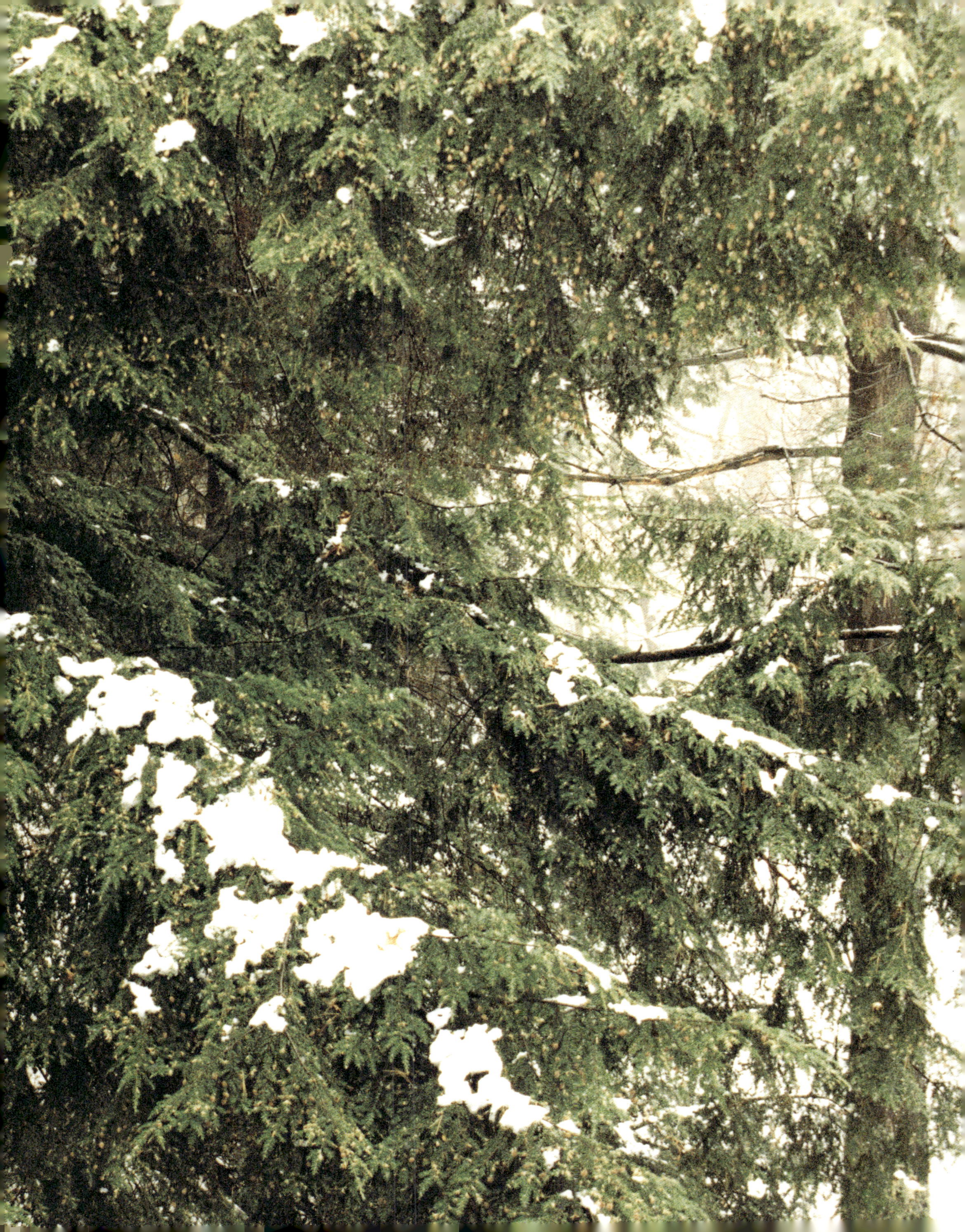

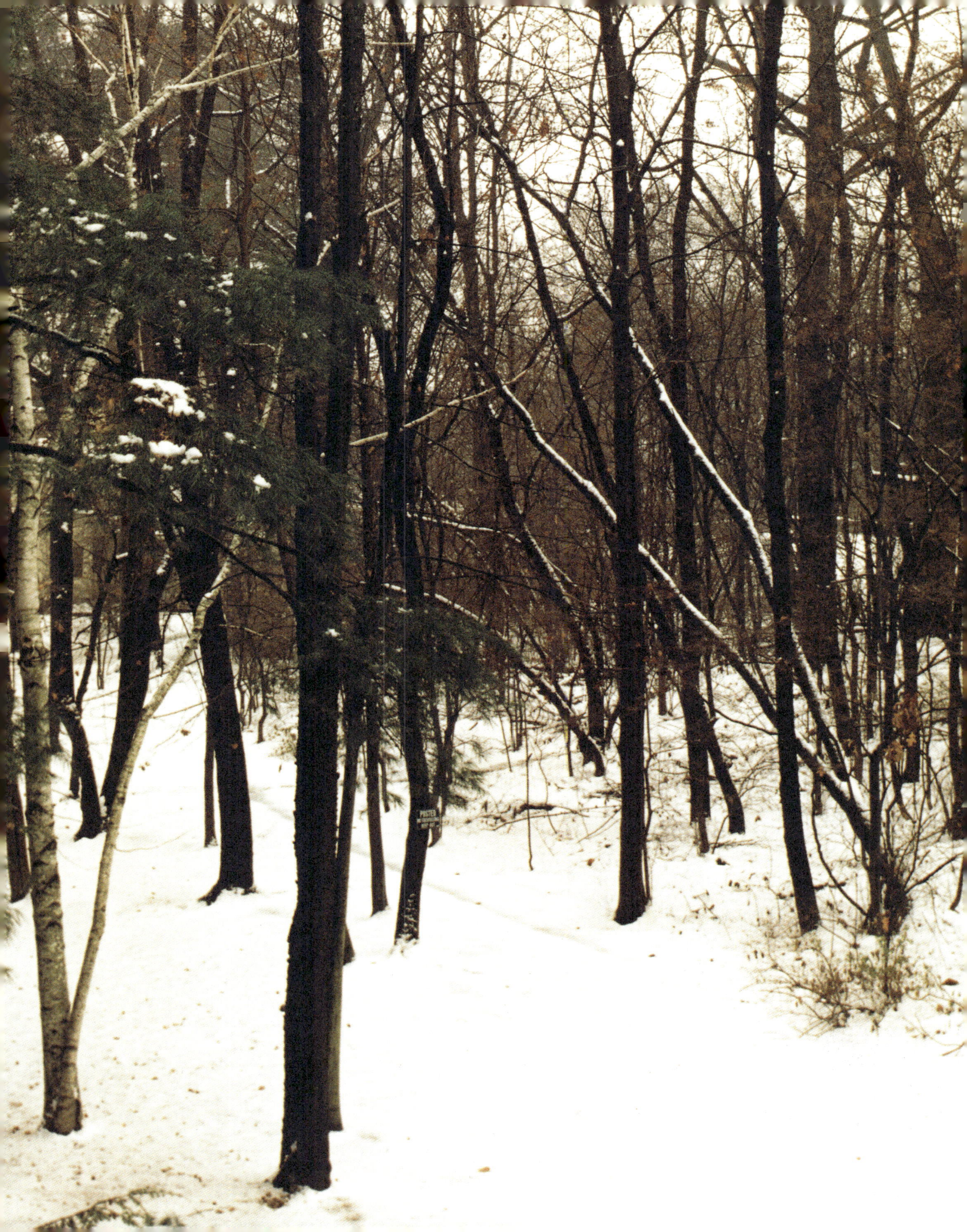

KEY AND CUE, NO. 351, 1994

KEY AND CUE, NO. 1060, 1994

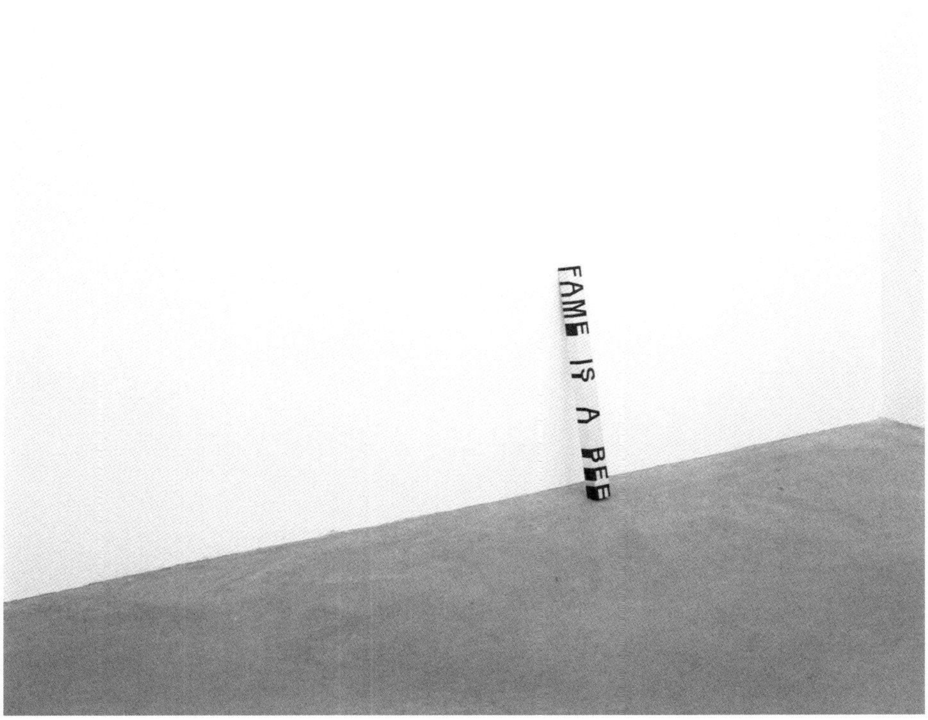

KEY AND CUE, NO. 1763, 1994

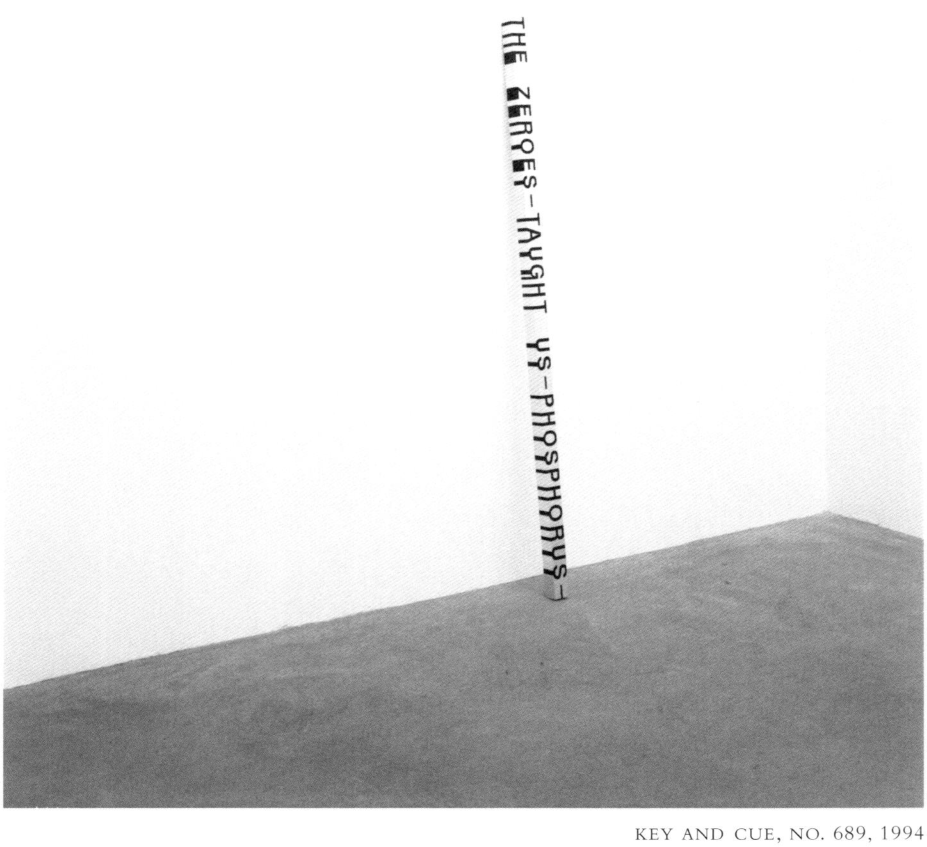

KEY AND CUE, NO. 689, 1994

Following pages—
UNTITLED (GUN), 1994
HOW DICKINSON STAYED HOME, 1992

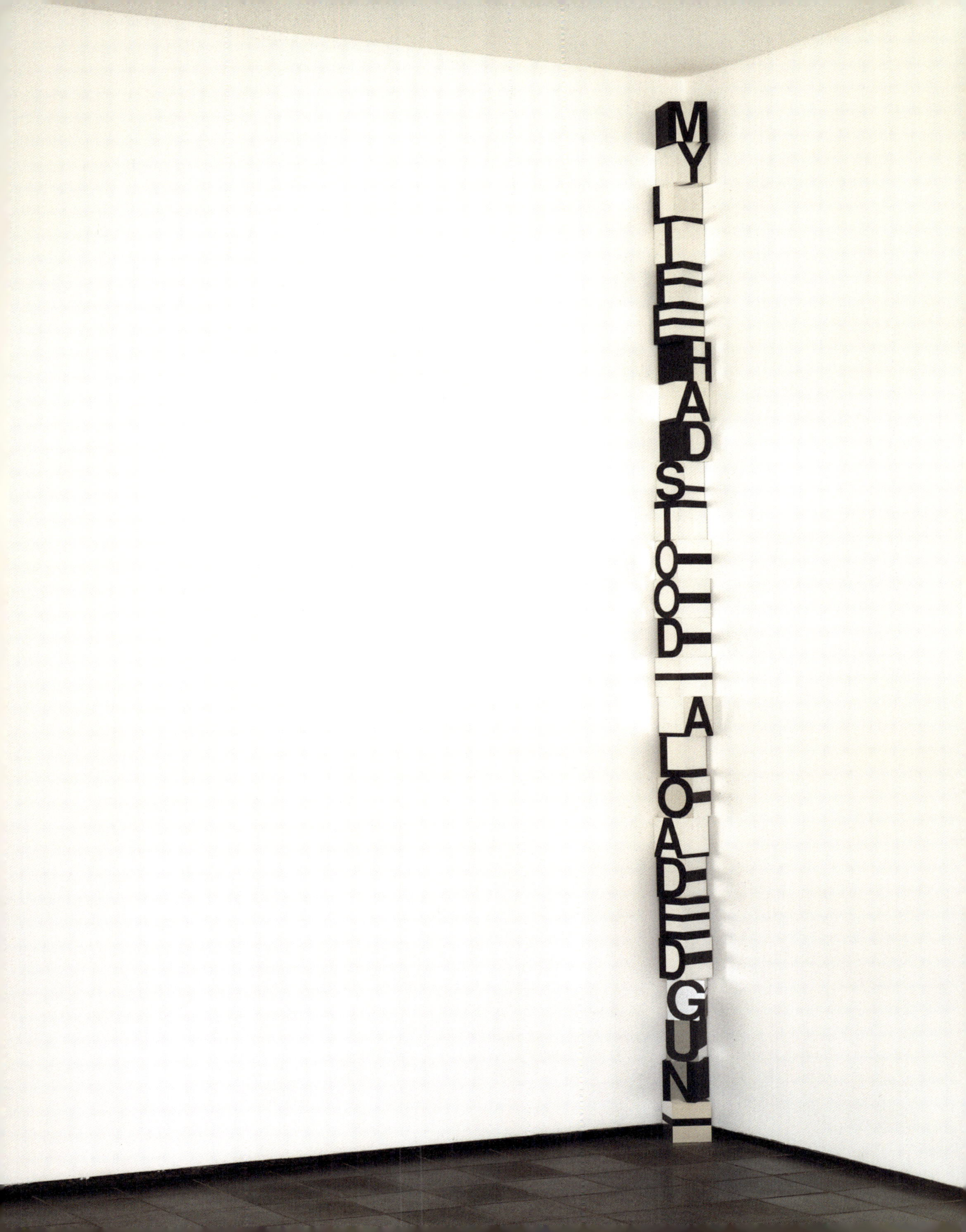

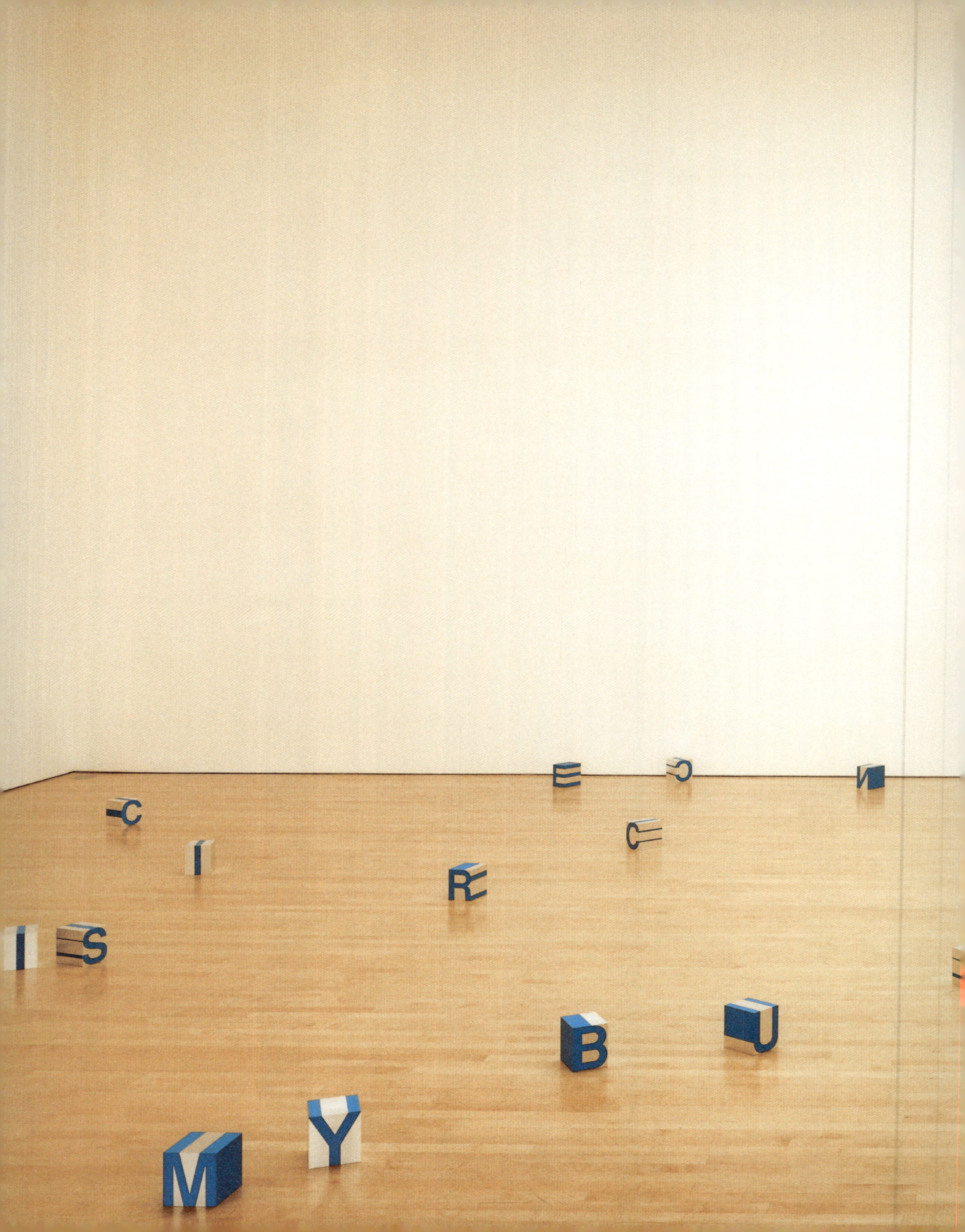

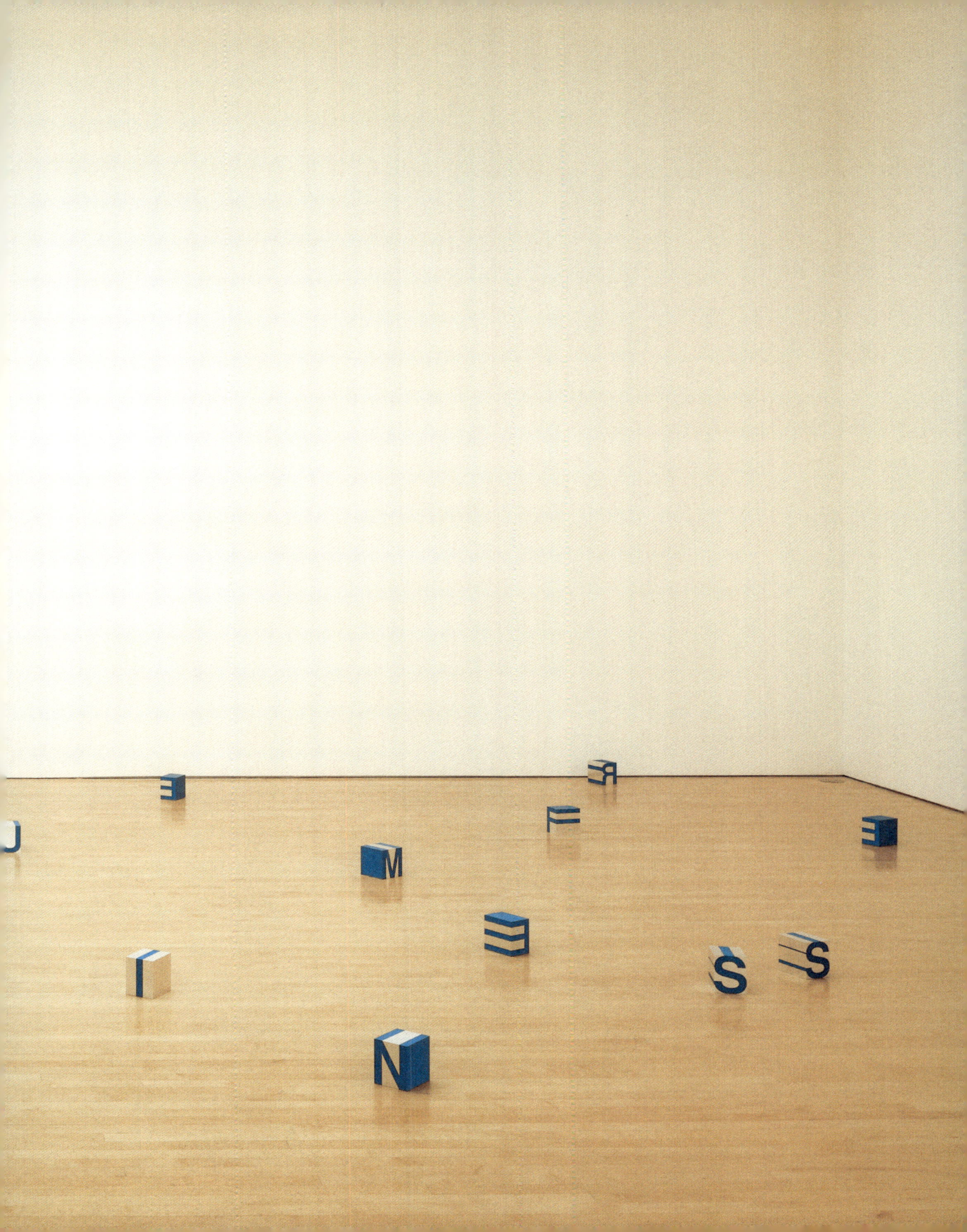

Following pages—
Manuscripts of Emily Dickinson's poems:
NO. 352, NO. 641, NO. 886, NO. 974, NO. 1027
(Reproduced actual size)

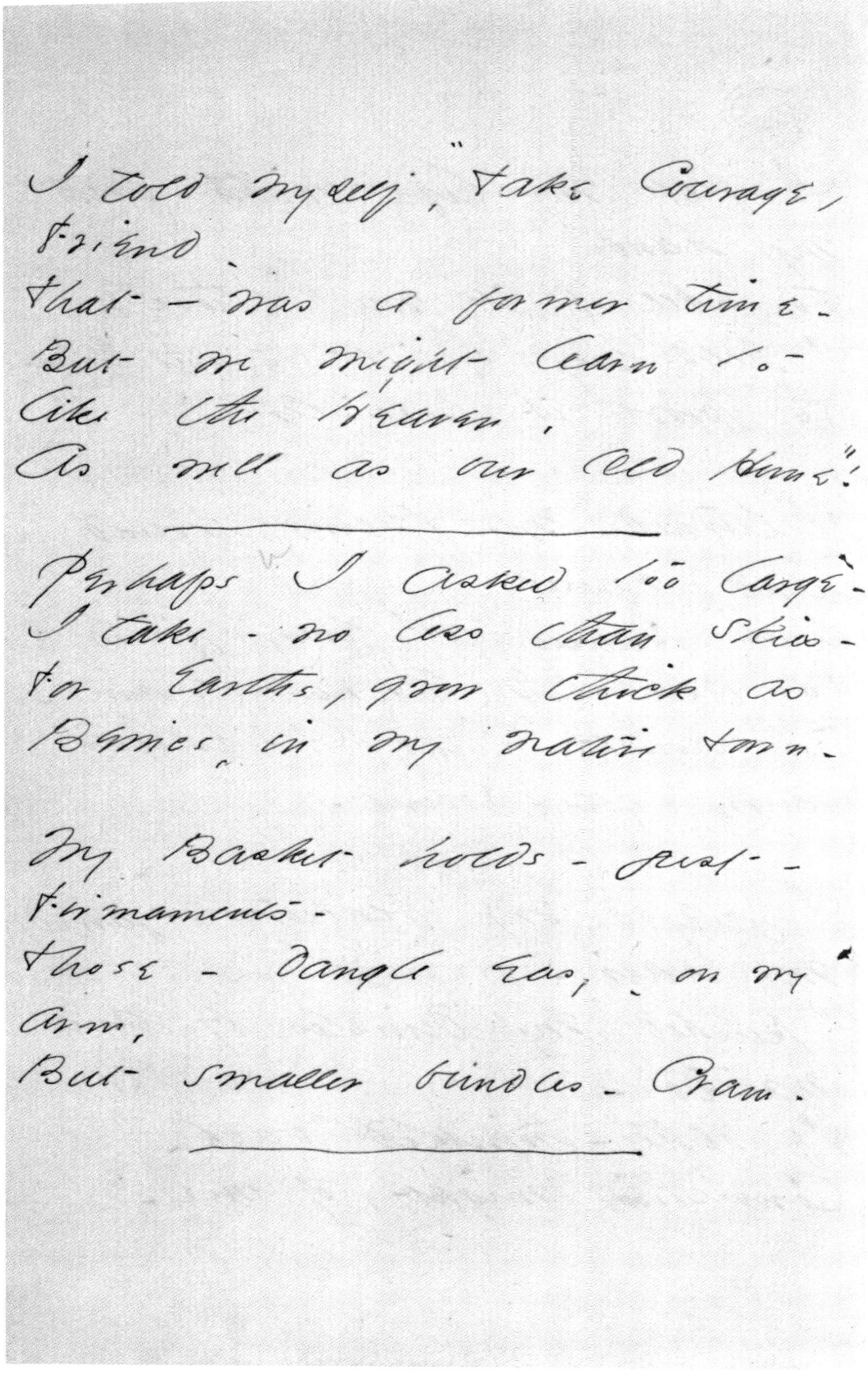
I told myself, "Take Courage,
Friend
That — was a former time —
But we might learn to —
like the Heaven,
As well as our Old Home!"

Perhaps I asked too large —
I take — no less than skies —
For Earths, grow thick as
Berries, in my native town —

My Basket holds — just —
Firmaments —
Those — dangle easy, on my
Arm,
But smaller bundles — Cram.

Size Circumscribes - it has
no room
For petty furniture -
The Giant $^{+}$tolerates no Gnat
For $^{+}$Ease of Gianture -

Repudiates it, all the more -
Because intrinsic size
$^{+}$Ignores the possibility
Of $^{+}$Calumnies - or Flies.

+ Simple Gianture - Because
of Gianture + Excludes + Jealousies.
+ Entertains

(41)

These tested Our Horizon -
Then disappeared
As Birds before achieving
A Latitude.

Our Retrospection of
Them
A fixed Delight,
But Our Anticipation
A Dice - a Doubt -

0 yes –

The Soul's distinct
Connection
With immortality
Is best disclosed by
Danger
Or quick Calamity –

As Lightning on a
Landscape
+ Exhibits Sheets of Place –
+ Not yet suspected – but
for + Flash –
And + Click – and Suddenness.

+ developes + still unsus-
pected + Fork + Bolt –

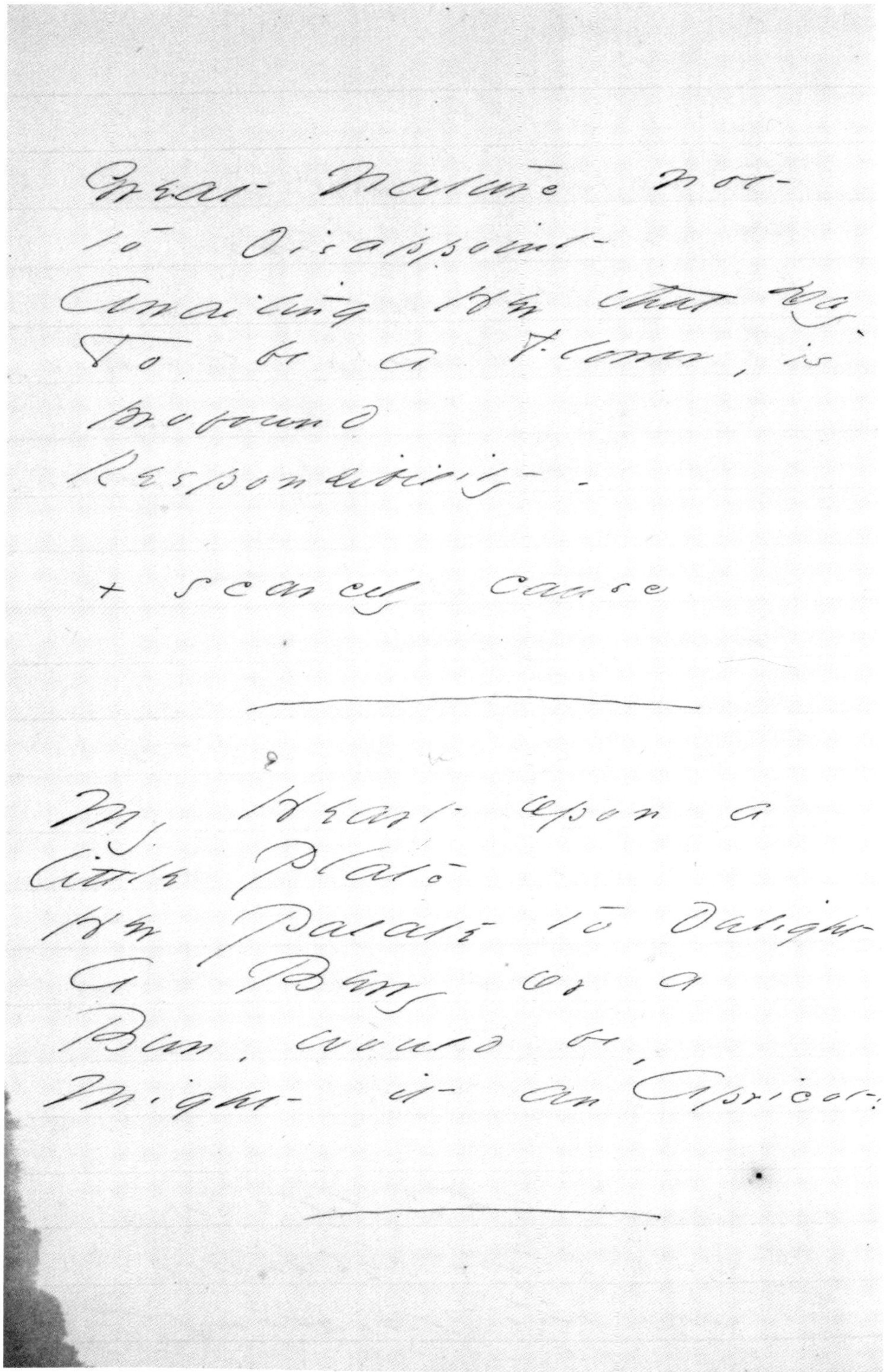
Great Nature not
to disappoint
Awaiting Her that Day -
To be a Flower, is
profound
Responsibility -

+ scarcely cause

My Heart upon a
little Plate
Her palate to delight
A Berry or a
Bun, would be,
Might it an Apricot -

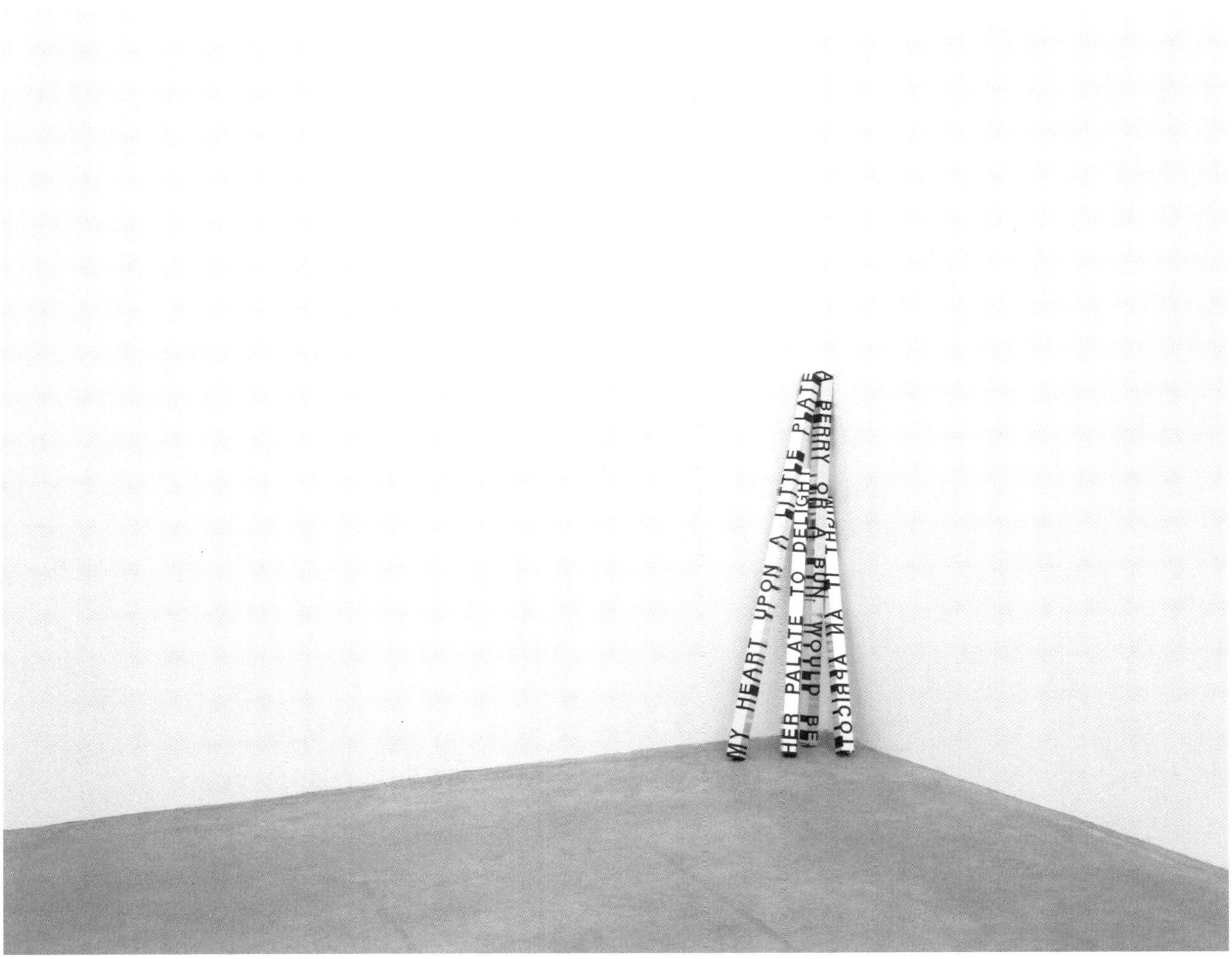

WHEN DICKINSON SHUT HER EYES, NO. 1027, 1993

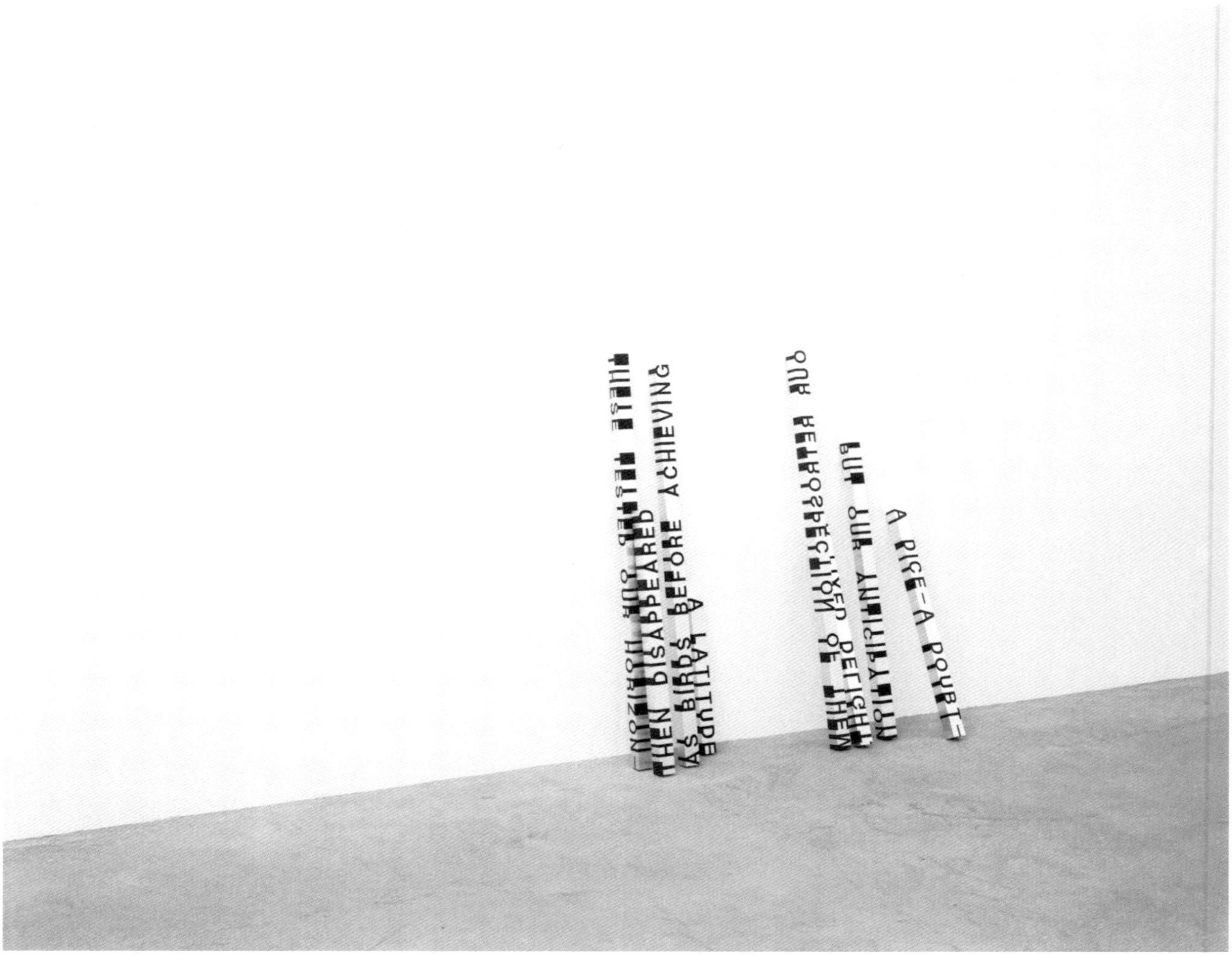

WHEN DICKINSON SHUT HER EYES, NO. 886, 1993

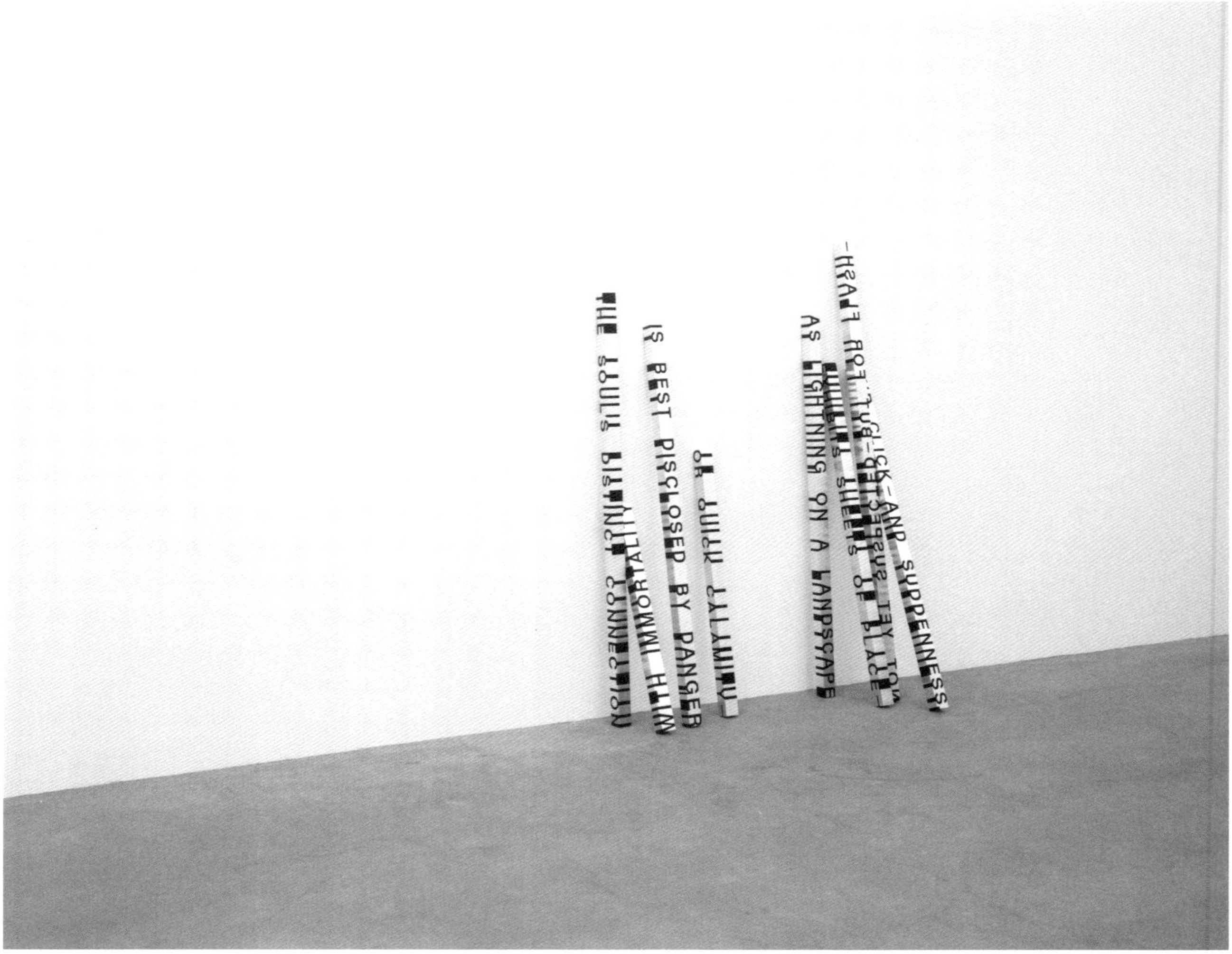

WHEN DICKINSON SHUT HER EYES, NO. 974, 1993

Earths Grow Thick

Cover July 1995, view from the window of Emily Dickinson's bedroom looking west.

6-15 Views from the windows of Emily Dickinson's bedroom at 280 Main Street, Amherst, Massachusetts: looking west towards *Evergreen* and south, across Main Street. Photographs taken December 1994.

17-22 *KEYS AND CUES*, 1994.
Solid aluminum and plastic. 2 x 2" by variable length from 27 to 103". Each *Key and Cue* is a quote of the first line of a poem by Emily Dickinson. Courtesy of Matthew Marks Gallery, New York and Jablonka Galerie, Cologne.
Installations: Texas Gallery, Houston, 1994; Susanne Hilberry Gallery, Birmingham, Michigan, 1995.
Collections: No. 153—Private collection, Washington D. C.; *No. 351*—Private collection, Bern, Switzerland; *No. 465*—Mr. and Mrs. Alexander Stuart, Lake Forest Illinois; *No. 533*—Private collection, Germany; *No. 813*—Michael A. Caddell and Tracey D. Conwell, Houston; *No. 825*—Mr. and Mrs. Max Herzstein, Houston; *No. 1182*—Private collection, Ann Arbor; *No. 1340*—Helen Elizabeth Hill Trust, Houston; *No. 1755*—Susan and Roy O'Conner, Missoula, Montana.

17 *Key and Cue, No. 351:*
I FELT MY LIFE WITH BOTH MY HANDS

Key and Cue, No. 825:
AN HOUR IS A SEA

18 *Key and Cue, No. 1060*:
AIR HAS NO RESIDENCE, NO NEIGHBOR

Key and Cue, No. 983:
IDEALS ARE THE FAIRY OIL

21 *Key and Cue, No. 1763*:
FAME IS A BEE

Key and Cue, No. 1158:
BEST WITCHCRAFT IS GEOMETRY

22 *Key and Cue, No. 689*:
THE ZEROES–TAUGHT US–PHOSPHORUS–

Key and Cue, No. 1182:
REMEMBRANCE HAS A REAR AND FRONT–

Key and Cue, No. 153:
DUST IS THE ONLY SECRET–

Key and Cue, No. 1340:
A RAT SURRENDERED HERE

Key and Cue, No.465:
I HEARD A FLY BUZZ–WHEN I DIED–

Key and Cue, No. 1755:
TO MAKE A PRAIRIE IT TAKES A CLOVER AND ONE BEE

Key and Cue, No. 533:
TWO BUTTERFLIES WENT OUT AT NOON–

Key and Cue, No. 632:
THE BRAIN–IS WIDER THAN THE SKY–

Key and Cue, No. 889:
CRISIS IS A HAIR

Key and Cue, No. 813:
THIS QUIET DUST WAS GENTLEMEN AND LADIES

24-25 *UNTITLED (GUN),* 1994.
Solid aluminum and plastic. 26 units each 5 x 5" by variable width. Stacked height: 10' 10". The text: MY LIFE HAD STOOD–A LOADED GUN– is from poem No. 754 by Emily Dickinson. Courtesy of Matthew Marks Gallery, New York and Jablonka Galerie, Cologne.
Installations: Jablonka Galerie, Cologne.

26-27 *HOW DICKINSON STAYED HOME,* 1992-93.
Solid aluminum and plastic. 25 units each 5 x 5" by variable length. The text:
MY BUSINESS IS CIRCUMFERENCE
is from a letter by Emily Dickinson.
Installations: Margo Leavin Gallery, Los Angeles; 1993; Kunsthalle Basel, "21st Century," 1993; The Museum of Contemporary Art, Los Angeles, 1994; The St. Louis Art Museum, 1995; Sammlung Goetz, Munich, 1995.
Collections: The Museum of Modern Art, New York; The Museum of Contemporary Art, Los Angeles; Sammlung Goetz, Munich.

29-32 Manuscripts of poems by Emily Dickinson (Reproduced actual size).

29 *No. 352*
"Perhaps I asked too large–"

30 *No. 641*
"Size circumscribes–it has no room"

31 *No. 886:*
"These tested Our Horizon–"

32 *No. 974*
"The Soul's distinct connection"

33 *No. 1027:*
"My Heart upon a little Plate"

35-38 *WHEN DICKINSON SHUT HER EYES,* 1993.
Solid aluminum and plastic. 2 x 2" by variable length. Each work reproduces a complete poem by Emily Dickinson.
Courtesy of Matthew Marks Gallery, New York and Jablonka Galerie, Cologne.
Installations: Mary Boone Gallery, New York, 1993; De Pont Foundation for Contemporary Art, Tilburg, The Netherlands, 1994; "Das Américas," Galerie Luisa Strina and Instituto Goethe, São Paulo, 1994; "In a Different Light," University of California at Berkeley, 1995; "Currents 61: Roni Horn," The St. Louis Art Museum, 1995.
Collections: No. 1027: Janet de Botton, London; *No. 352*: Donald Bryant, St. Louis; *No. 912:* Private collection, São Paulo.

35 *No. 1027*
MY HEART UPON A LITTLE PLATE
HER PALATE TO DELIGHT
A BERRY OR A BUN, WOULD BE,
MIGHT IT AN APRICOT!

36 *No. 886*

THESE TESTED OUR HORIZON—
THEN DISAPPEARED
AS BIRDS BEFORE ACHIEVING
A LATITUDE.

OUR RETROSPECTION OF THEM
A FIXED DELIGHT,
BUT OUR ANTICIPATION
A DICE—A DOUBT—

No. 641

SIZE CIRCUMSCRIBES—IT HAS NO ROOM
FOR PETTY FURNITURE—
THE GIANT TOLERATES NO GNAT
FOR EASE OF GIANTURE—

REPUDIATES IT, ALL THE MORE—
BECAUSE INTRINSIC SIZE
IGNORES THE POSSIBILITY
OF CALUMNIES—OR FLIES.

38 *No. 974*

THE SOUL'S DISTINCT CONNECTION
WITH IMMORTALITY
IS BEST DISCLOSED BY DANGER
OR QUICK CALAMITY—

AS LIGHTNING ON A LANDSCAPE
EXHIBITS SHEETS OF PLACE—
NOT YET SUSPECTED—BUT FOR FLASH—
AND CLICK—AND SUDDENNESS.

No. 912

PEACE IS A FICTION OF OUR FAITH—
THE BELLS A WINTER NIGHT
BEARING THE NEIGHBOR OUT OF SOUND
THAT NEVER DID ALIGHT.

No. 352

PERHAPS I ASKED TOO LARGE—
I TAKE—NO LESS THAN SKIES—
FOR EARTHS, GROW THICK AS
BERRIES, IN MY NATIVE TOWN—

MY BASKET HOLDS—JUST—FIRMAMENTS—
THOSE—DANGLE EASY—ON MY ARM,
BUT SMALLER BUNDLES—CRAM.

46-55 Views from the windows of Emily Dickinson's bedroom at 280 Main Street, Amherst, Massachusetts: looking west towards *Evergreen* and south, across Main Street. Photographs taken July 1995.

103 An oak tree on the Dickinson property at 280 Main Street, Amherst, Massachusetts, 1995.

Selected Bibliography

MONOGRAPHS

Diacono, Mario. "Unique Forms of Deviation in Space." Boston: *Mario Diacono*, October 1988.

Friedel, Helmut; Horn, Luise. *Roni Horn*. Munich: Kunstraum, 1983.

Fuchs, Rudi; Kersting, Hannelore. *Things Which Happen Again*. Mönchengladbach and Münster: Städtisches Museum Abteiberg and Westfälischer Kunstverein, 1991.

Gilbert-Rolfe, Jeremy. *Roni Horn: Pair Objects I, II and III*. Paris and New York: Galerie Lelong, 1988.

Gonzalez-Torres, Felix; Horn, Roni; et al. *Felix Gonalez-Torres, Roni Horn*. Munich: Sammlung Goetz, 1995.

hooks, bell; Gonzalez-Torres, Felix; et. al. *Earths Grow Thick*. Columbus, Ohio: Wexner Center for the Arts, 1996.

Howard, Jan; Horn, Roni. *Inner Geography*. The Baltimore Museum of Art, 1994.

Johnson, Ken. "Material Metaphors." New York: *Art in America* (February 1994): 70-79 and cover.

Kellein, Thomas; Horn, Roni. *Making Being Here Enough*. Basel and Hannover: Kunsthalle Basel and Kestner-Gesellschaft, 1995.

Kern, Hermann. *Roni Horn*. Munich: Kunstraum, 1980.

Kertess, Klaus. *Roni Horn: Surface Matters*. Los Angles: The Museum of Contemporary Art, 1990.

Koepplin, Dieter. *Roni Horn*. Basel: Museum für Gegenwartskunst, 1995.

O'Rourke, Meg. "The Weight of The Word." New York: *Arts Magazine* (November 1991): 58-60.

Schwarz, Dieter. *Rare Spellings, Selected Drawings: 1985-92*. Düsseldorf: Richter Verlag, 1993.

Spector, Nancy. *Roni Horn: Being Double*. Tilburg, The Netherlands: De Pont Foundation for Contemporary Art, 1994.

Spinelli, Claudia. "A Talk with Roni Horn." Zurich: *Das Kunst-Bulletin* (No. 10, October, 1995): 10-19, and cover.

Tillman, Lynne. *Gurgles, Sucks, Echoes.* Cologne and New York: Jablonka Galerie, Matthew Marks Gallery, 1995.

Wylie, Charles. *Currents 61: Roni Horn*. The St. Louis Art Museum, 1995.

BOOKS BY THE ARTIST

To Place: Bluff Life. 1990—36pp. 14 color reproductions; Edition 1150; Special Edition 150; Peter Blum Edition, New York.

To Place: Folds. 1991—72pp. 36 color images; letterpress throughout; Edition 700; Mary Boone Gallery, New York.

To Place: Lava. 1992—96pp. 16 color and 29 tritone images; letterpress throughout; Edition 750; Special Edition: 100; Roni Horn, New York.

To Place: Pooling Waters. 1994—Volume 1: 96pp. 27 color and 25 duotone images—Volume 2: 176pp. 4 images; Text—English/Icelandic; Edition 1000; Special Edition 6; Verlag der Buchhandlung Walther König, Cologne.

"Verne's Journey." (Insert). Zurich: *Parkett*, (No. 39, 1994): 131-143.

To Place: Verne's Journey, 1995—56pp. 13 color and 8 duotone images; Edition 1000; Verlag der Buchhandlung Walther König, Cologne.

FORTHCOMING BOOK: *To Place: Haraldsdóttir*, 1996—96pp. 30 color and 31 duotone images; Edition 1000.

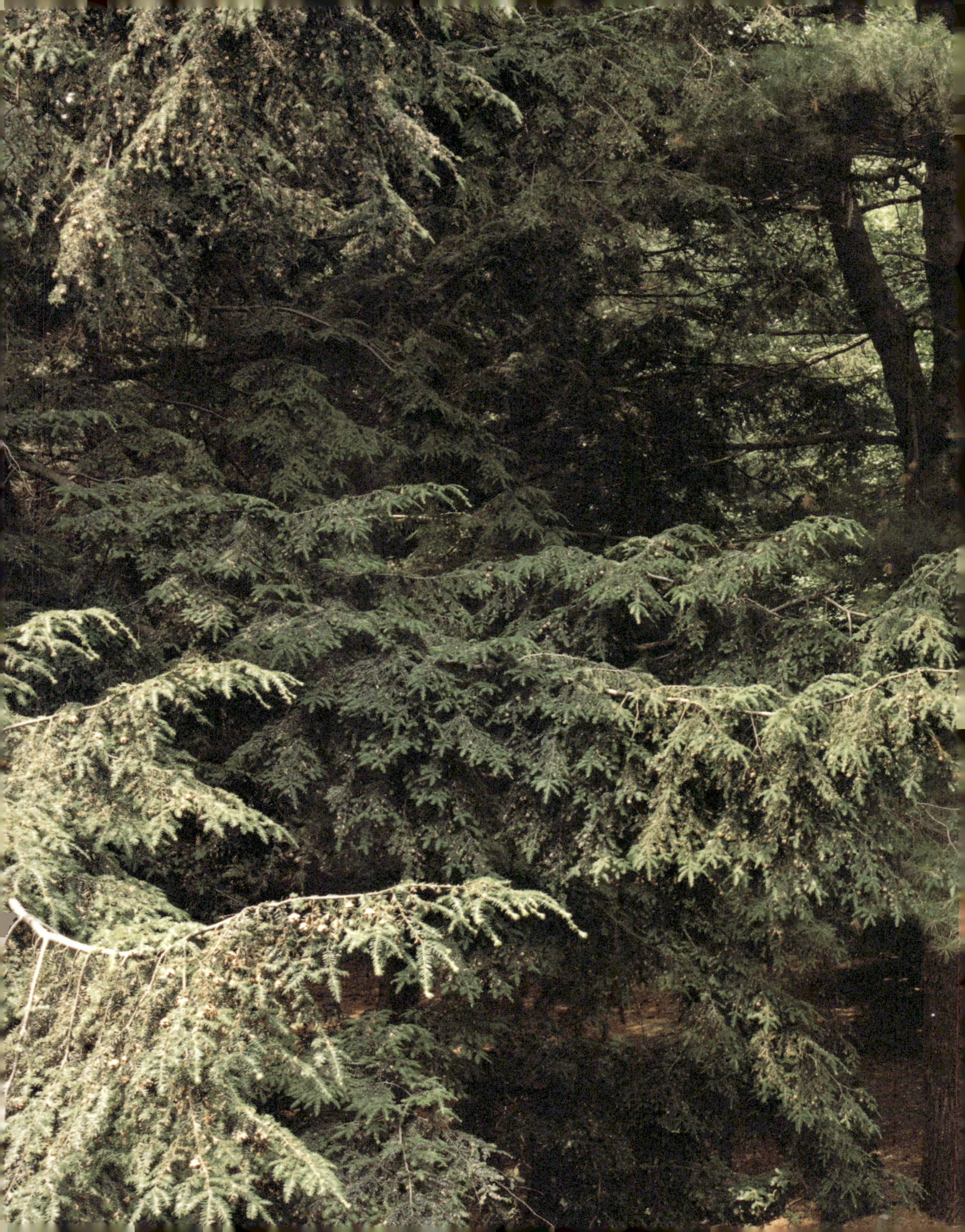

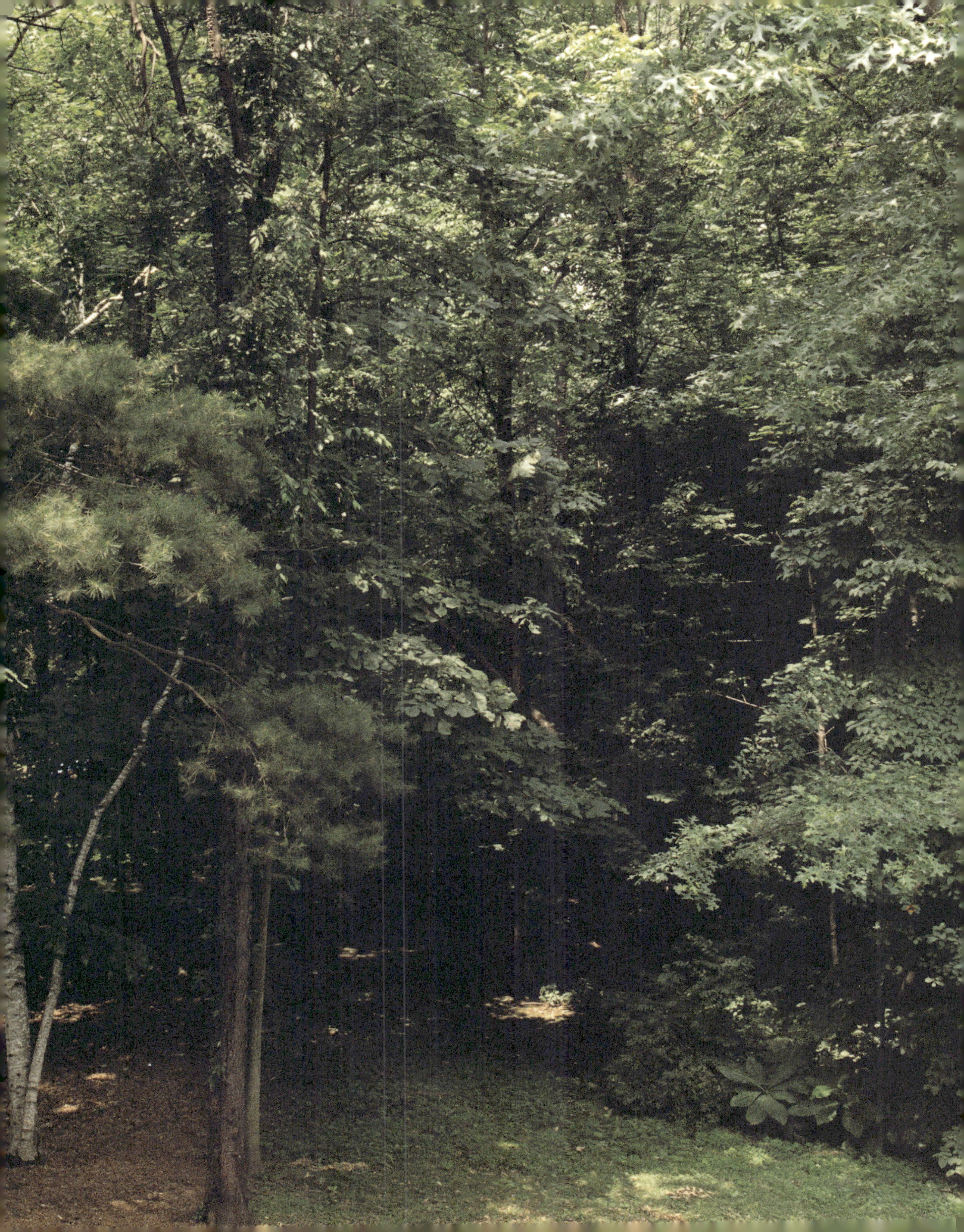

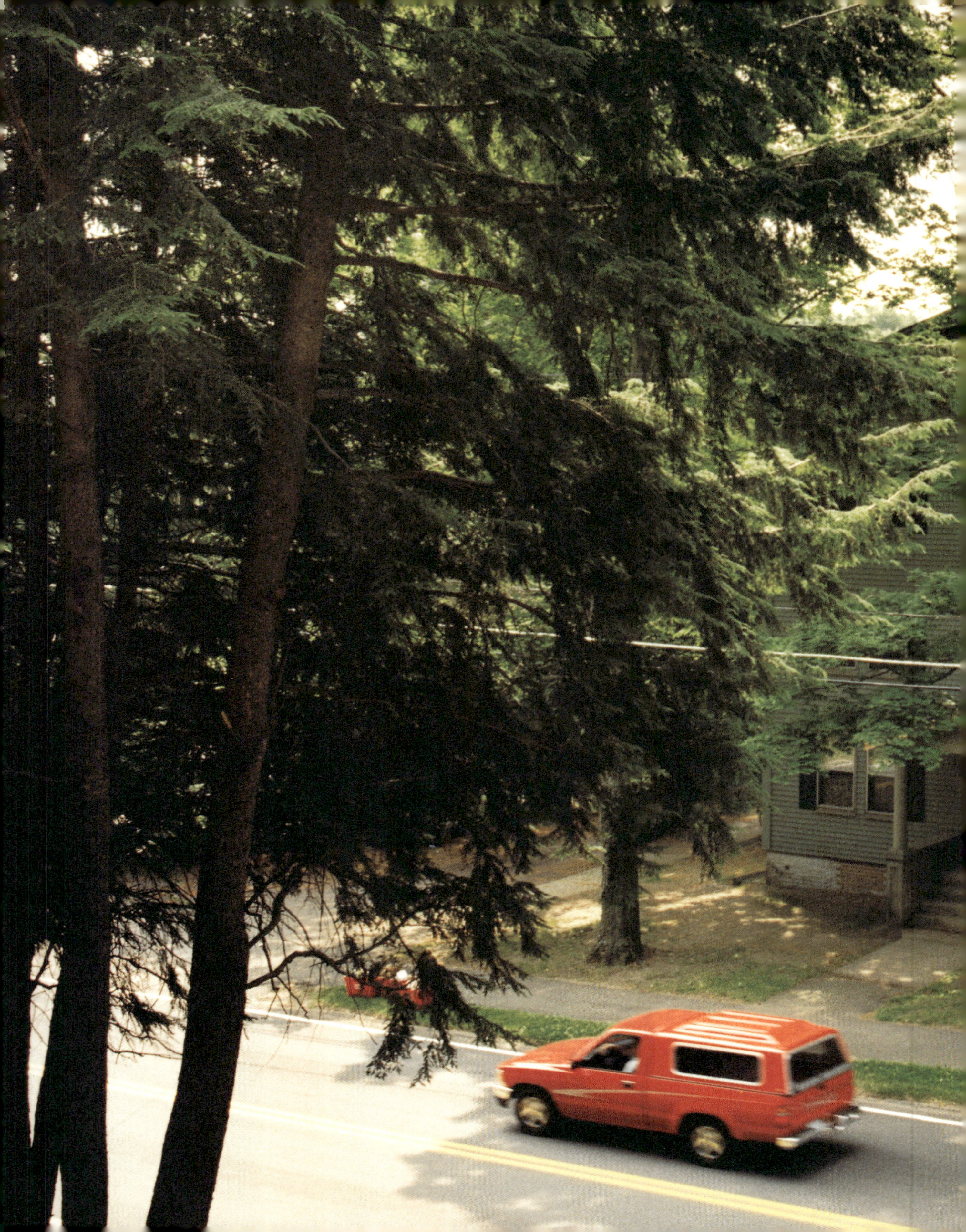

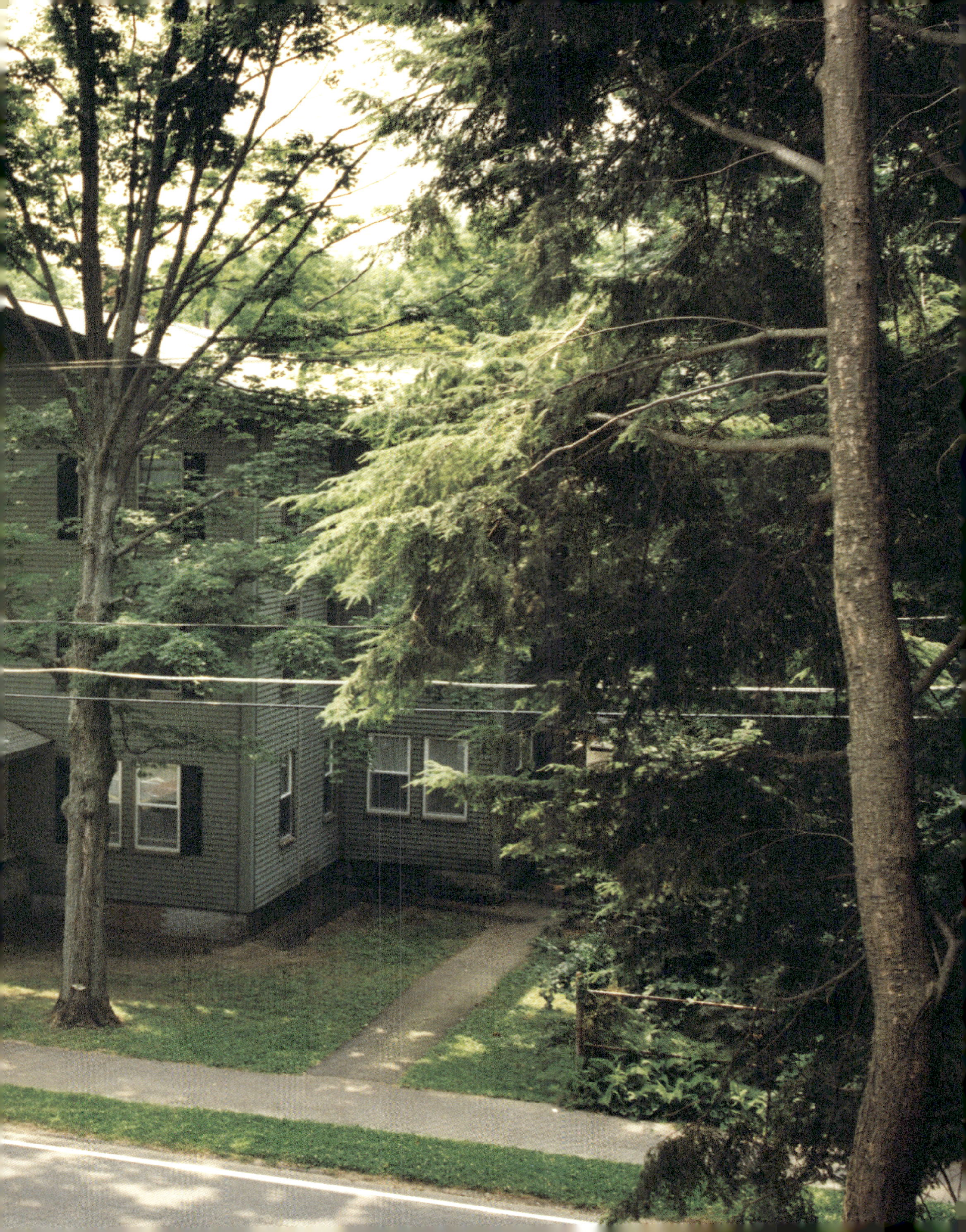

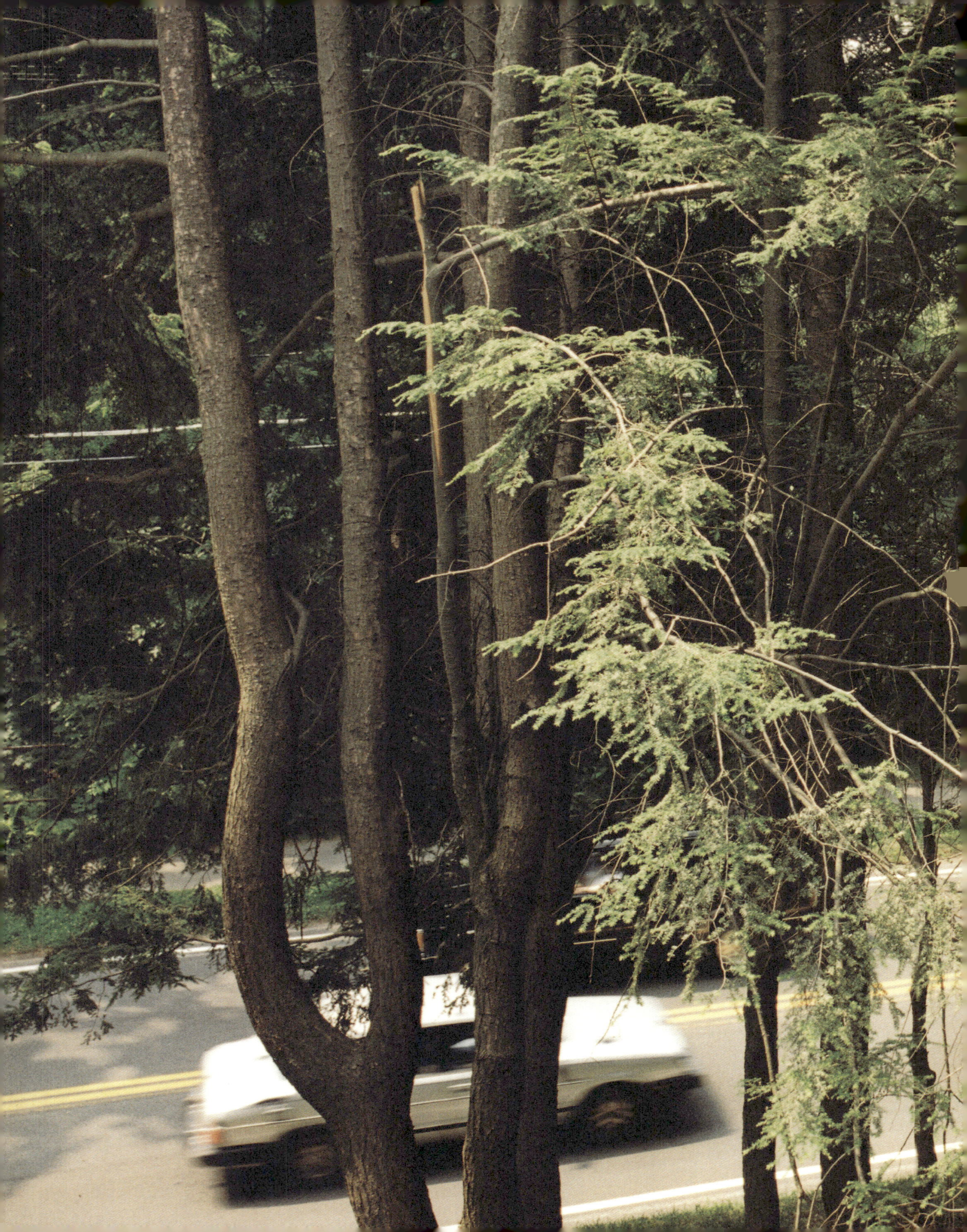

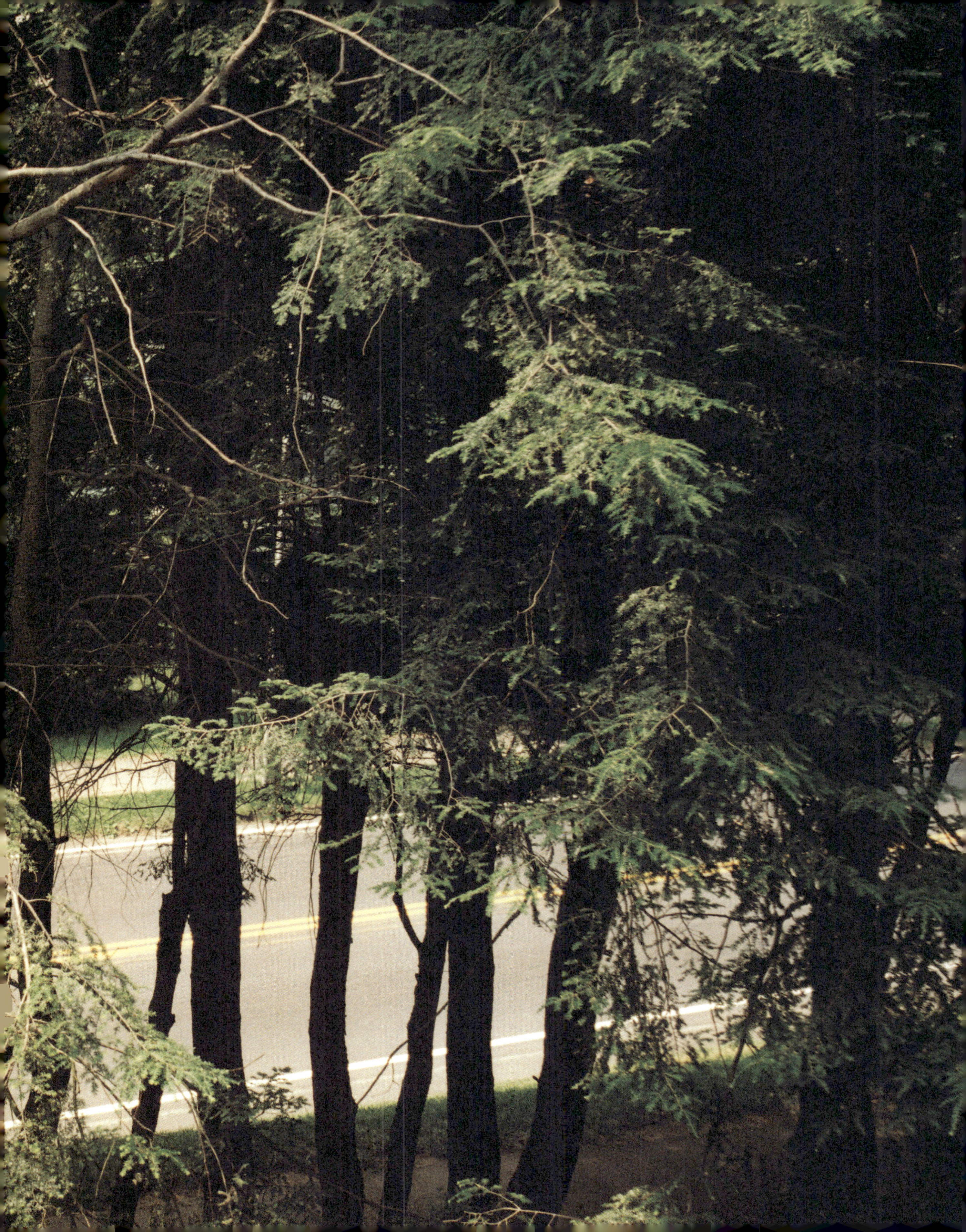

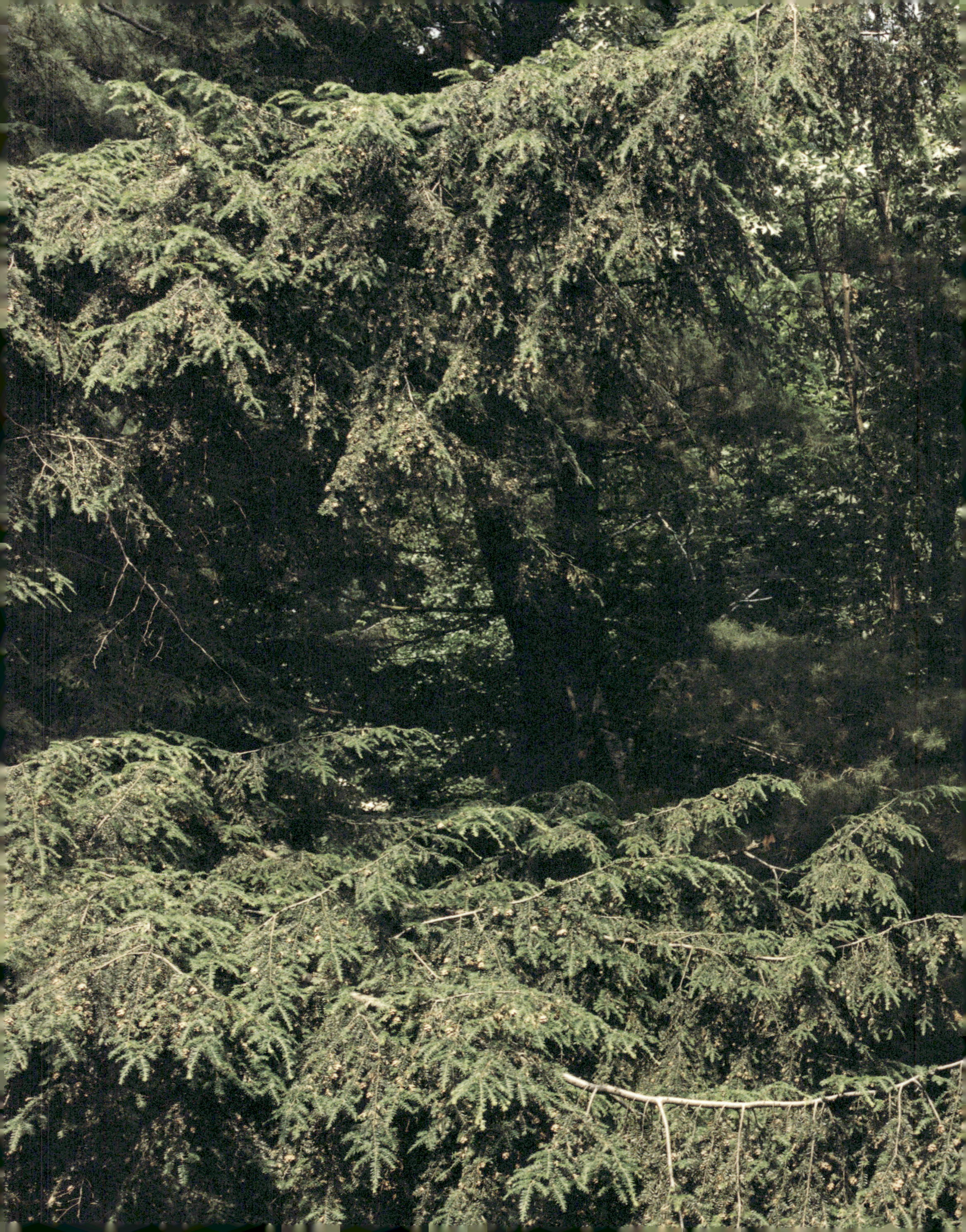

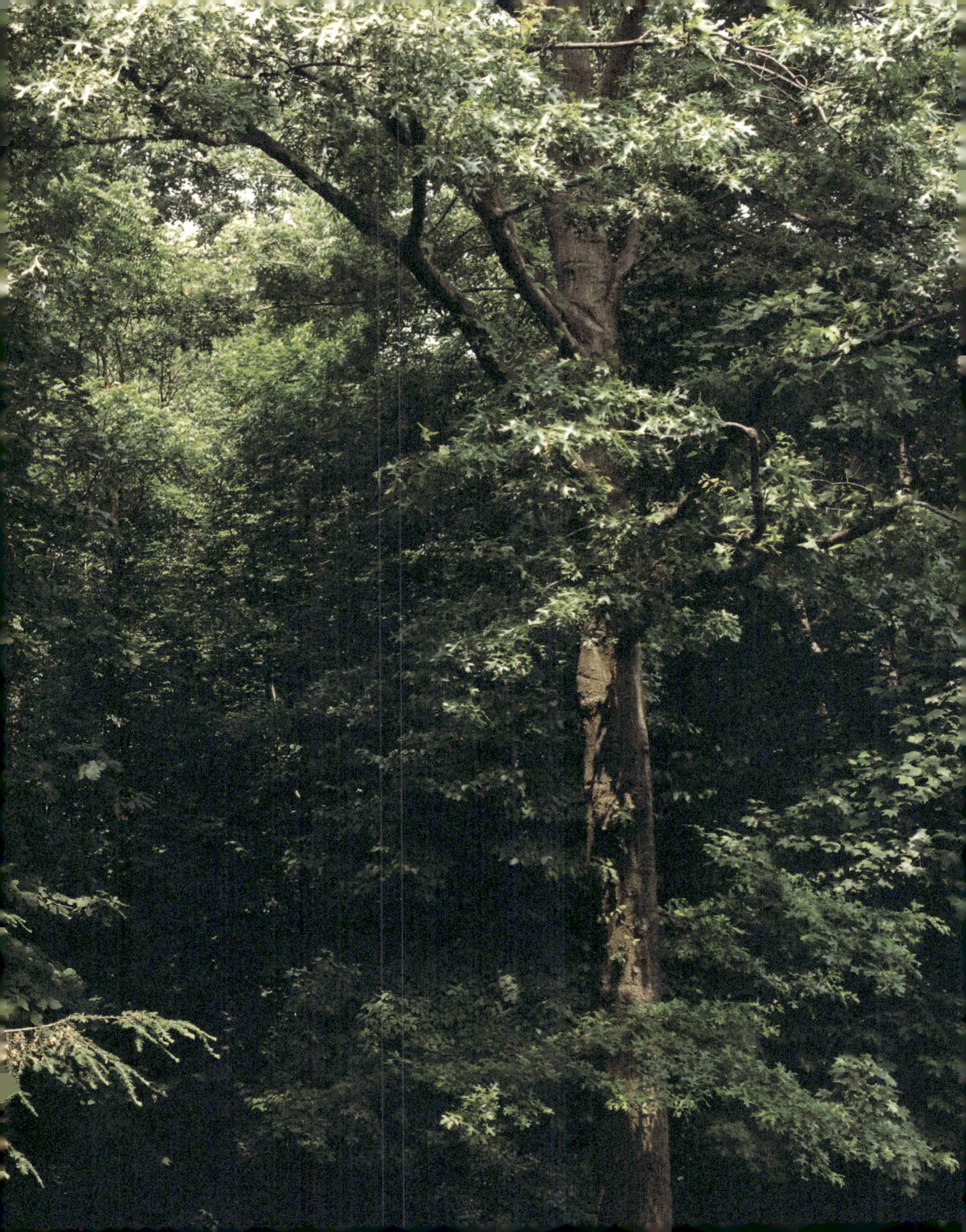

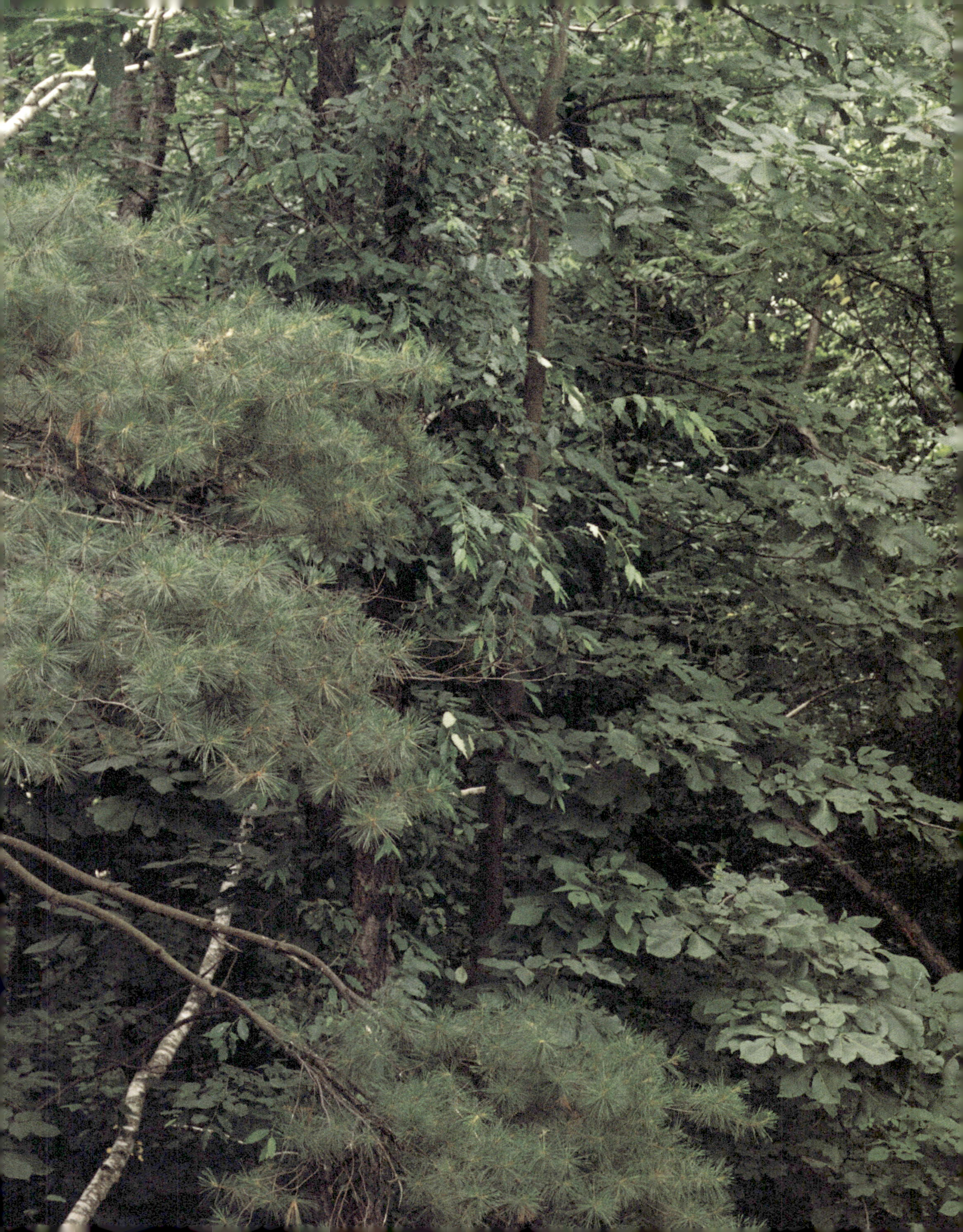

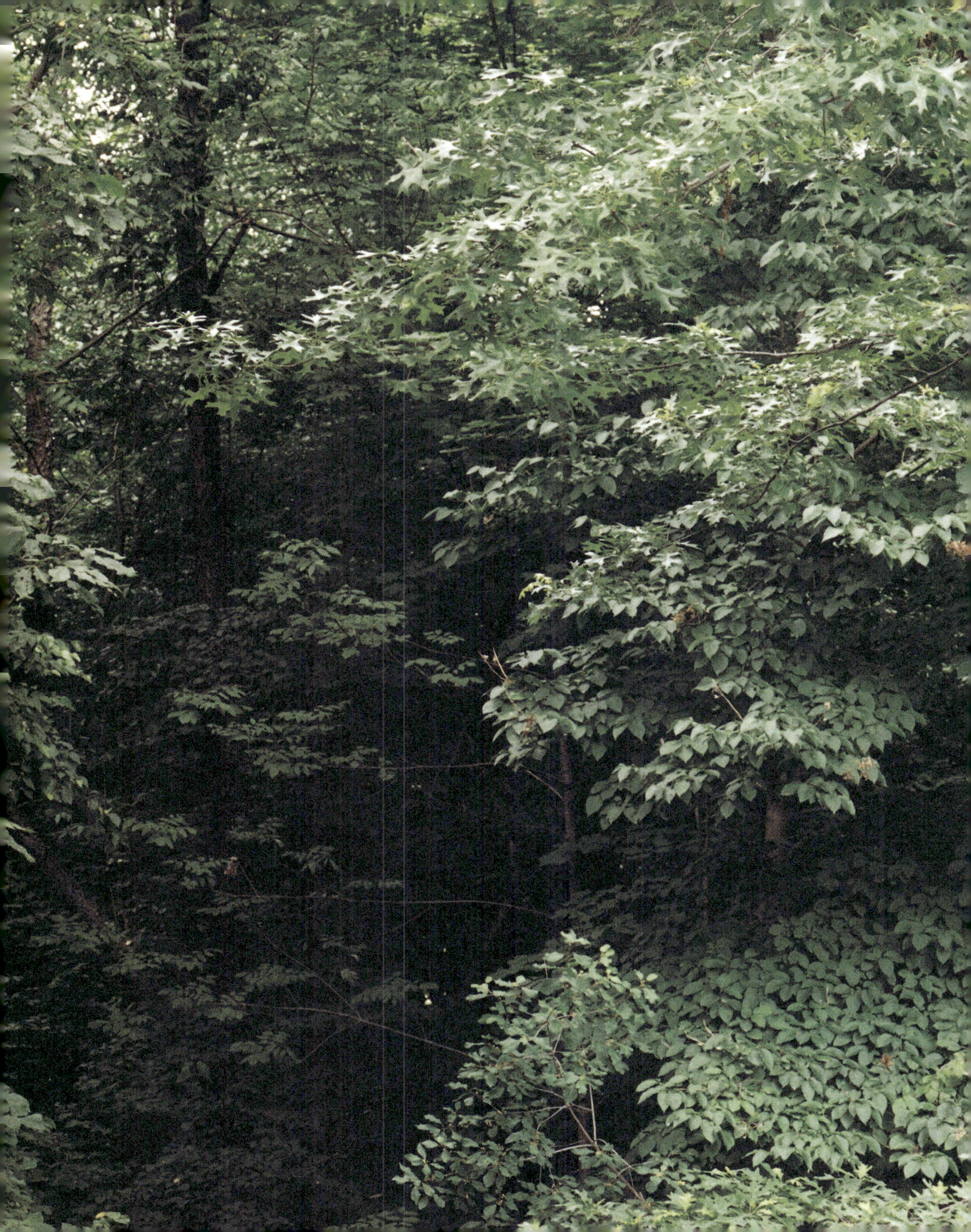

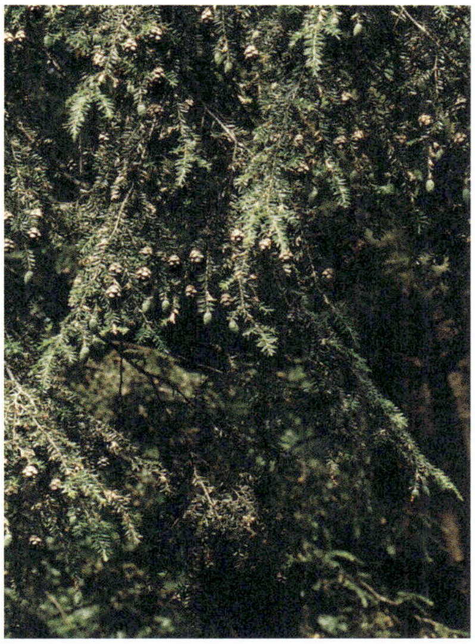

between us: traces of love— Dickinson, Horn, Hooks

bell hooks

When I imagine angels they are not some winged fantastical apparitions hanging on the rafters of the soul looking down. They are for me always embodied concrete guardians, those that the Ethiopian Copts called "zars." In the images the Copts painted to give face to their beliefs angels are human-like. Angels were only useful in my childhood because they were guardians—there to see to the fulfillment of one's need, there to protect. I always chose my angels; sometimes they were real human beings and sometimes not. The angel of my solitary spirit, the one who guided me through poetry to the contemplative, to the mystical understanding that one could know the Divine through direct experience and encounter—this angel was Emily Dickinson.

Emily D., as I called her when I was ten years old, was certain of one thing—that solitude was essential for the nurturing of her creative imagination. Solitude was the space where her soul could come out of hiding and be heard. To talk with the world face to face, in the ordinary business of going out of one's house to shop, to meet—all of these disturbances would have sent the soul marooned to some place of forgetfulness. It would have been harder to find its intimacy again.

There was for Emily D. no better place to care for the soul than inside the home—the house of one's actual concrete daily living. Writing to her brother Austin in 1851 she shared: "Home is a holy thing—nothing of doubt or distrust can enter its blessed portals...here seems indeed to be a bit of Eden which not the sin of any can utterly destroy." In the Eden that was her home Emily Dickinson could subvert and defy all the restraints the world might have placed on her imagination and her being had she chosen to venture out more, had she chosen to live the normal life of a white woman of her times. Although I never thought of Emily D. in my childhood as white. I must do so now—to understand the luxury of her solitude, its privilege. No black woman of her day and time could have lived such a solitary life. In the world she lived in, black women were there to serve. These thoughts of race were never in my girlhood when Emily D. my sweet one, guardian and confidante, angel of my solitary spirit spoke to me. Sharing the magic of stillness and contemplation, she spoke to me most clearly and directly in the shadowy spaces behind her words.

As a girl the only information about Emily D. we learned in school was that she rarely came out of the house. Her solitude was presented to us as weird, as a mark of her strangeness. It was the space behind the poems that offered the insight that there was more to this solitude, something deeper. To this poet who "dwelled in possibility" solitude was the place for magic and alchemy—the space wherein one could meet the Divine face to face.

When I read "This is my letter to the World / That never wrote to Me—" I heard both lamentation and liberty. Emily D. seemed to both mourn her estrangement from the things of this world while exulting in the visions seclusion and solitary confinement brought to her. That she mourns the very worldliness she has actively chosen to forego participation in to nurture her creative vision clearly shows concern and care for that world. To be at home in the world, Emily Dickinson needed to create a pure space where she could dream and dream without interruption, without needing to fear that the innocence of her vision would be corrupted. Without the distraction of worldliness, Dickinson could fully open the realm of her senses and sensually experience space both in the domestic household and in the natural world.

Living an ascetic, almost monastic, existence she could hear more, see more, of everything that was happening in nature. Dickinson's poems reveal her understanding of the interconnectedness of all living organisms. In Rosemary Reuther's ecofeminist theology of earth healing, *Gaia and God,* she asks: "How do we connect ourselves and the meaning of our lives to these worlds of the very small and the very big, standing in between the dancing void of energy that underlies the atomic structure of our bodies and the universe, whose galaxies, stretching over vast space and time, dwarf our histories: Even our bodies, despite the appearance of continuity over time are continually dying and being reborn in every second." For Emily D. this connection was made by making a sacrament of the everyday, by remaining close to nature. Close enough to write of the inner life of flowers:

My nosegays are for Captives—
Dim—long expectant eyes,
Fingers denied the plucking,
Patient till Paradise.

To such, if they should whisper
Of morning and the moor,
They bear no other errand

Or in "Forbidden fruit": " 'Heaven' is what I cannot reach / The Apple on the Tree—" Or when meditating on the meaning of storms in nature she writes "The Soul's Storm":

It struck me—every Day—
The Lightning was as new
As if the Cloud that instant slit
And let the Fire through.

In many poems Dickinson contemplates the ordinary visible things in our daily experiences—a bird, a flower, the next door neighbor's house—all of which become ways to arrive at a deeper transcendental meaning of the mystery and meaning of life.

Throughout my writing life, I have walked a creative path guided by the life and work of Emily Dickinson. Now as a mature woman and writer in retrospect I can see the myriad ways her commitment to solitude and contemplation deeply threaten patriarchal norms. That is why her solitariness is usually still presented to the minds of young girls and boys as something strange and weird. Even though time has accorded her an unchallenged place in the canon of "great" and serious American literature, the meaning of her solitude still remains suspect and unclear to many who read and appreciate her work. It is rarely presented as essential to the workings of her creativity in the same way that the commitment to uninterrupted time of Rainer Maria Rilke, another poet who enthralled me during my girlhood, is extolled as a sign of his devotion to his art, his willingness to lay down the ground of his worldly being to make that open space within wherein the poems could enter and the visionary voice speak. Dickinson's solitariness is essential for the cultivation of her creative passion. It was in that solitary space that she found her most intimate connection to the divine, to that quality of yearning for ecstasy that would lead her to write:

Wild Nights—Wild Nights!
Were I with thee
Wild Nights should be
Our luxury!

Dickinson's poetry is a celebration of mystical union with the divine. It is during those times that she is most "alone with the Alone" that her imagination takes flight and soars. It is that soaring in a world where there is no time and space that she, the angel of solitary spirits, finds us and stands watch over our creative spirit.

Working imaginatively to claim that space where we can embrace contemplation and solitude, contemporary artists look to the presence of Emily Dickinson to inspire. Our engagement with her work connects us to one another. My journeying with Emily Dickinson brings me to the work of Roni Horn. In the fundamentalist Christian ethos of my childhood supplicants would ask of the Divine to let "the breath of the Lord now breathe on me." This shared breath is symbolic of the passage of divine spirit from one being to another. Like mouth to mouth resuscitation it calls us back to life. Emily Dickinson's work has this life sustaining power.

Horn acts to give visual expression to that power. By creating objects that embody the metaphysics inherent in Dickinson's vision and in her words, she both extends that breath and expands it. She opens up Dickinson, creating with, through, and beyond her. In her deconstructive appropriation of Dickinson, Horn maps out a visual typography of the contemplative imagination at work. To Horn "mapping is a way of ordering things visually." In her typographic drawings and photographs there is a visual palimpsest that leads the eye back to the words underneath, the words and life of Dickinson that ground Horn's imaginative quest for a spatial frontier where solitude can reinvent itself within the postmodern context. Horn concentrates on Emily Dickinson's solitude, recognizing that this is the source of her genius and power. In the narrative *When Dickinson Shut Her Eyes* (1993) Horn reminds us that the poet lived "sequestered from the world" because she knew "that going out into the world hampered her ability to invent." While she appropriates Dickinson's metaphysics of contemplation and solitude, Horn offers a vision of journeying that invites us to realize the possibility of a solitude we invent in spite of lives lived in a constantly frenetic and busy world. To return us to the quality of stillness lyrically evoked in Dickinson's poetics, Horn gives us the evocative object, sculptures of solid aluminum and plastic that urge a re-encountering of that contemplative stance. The sculpture *Key and Cue* No. 351 (1994) stands straight, against a plain sparse backdrop a single declarative statement asserts: "I FELT MY LIFE WITH BOTH MY HANDS." Or the *Untitled (Gun)* piece (1994)—blocks stacked that compel us to look and hear again Dickinson's words: "MY LIFE HAD STOOD—A LOADED GUN—." These sculptures have a purity of presence that insists on a world where all that is necessary has been stripped away.

In her life and work Emily Dickinson stripped away the excess, the clutter, to make room for clarity. The meditative quality of her words resonates in the images Horn creates to contain and further illuminate that world. The closeness Horn feels with Emily Dickinson, one that defies time and space impressed itself on her imagination. Reading Dickinson's letters and poems she found in the work a place of recognition, a mirroring of her own inner landscape. A geography of the heart begins to imprint itself on Horn's imagination. She would enter more fully into the psychic spaces inhabited by Emily Dickinson to return to herself, and her present. Horn's work is not sentimental or nostalgic. It is a careful analytical imaginative mapping of an inner geography of soul—and the reaching through time of one artist toward another. Articulating this sense of space in "Anatomy and Geography," a text included in *Pooling Waters*, Horn shares the insight that: "Inner geography is a plain knowledge of oneself, a kind of common sense gathered through repeated exposure to distilling experiences. Inner geography maps peace of mind in the world as it is and not as I imagine it." Like Dickinson's, Horn's approach is that of the documentary observer, the surveyor who through a systematic process of looking allows the landscape to reveal itself without the imposition of a pre-determined view. She fixates on the landscape of Dickinson's imagination doubling this with her concern for an inner geography that transcends that landscape, that makes it over so that it becomes a new site, a new place. Both *When Dickinson Shut Her Eyes* (1993) and *How Dickinson Stayed Home* (1992-93) remind us that located at the site of this inner geography are multiple narratives, diverse points of entry. Unlike Dickinson who was always mapping an inner geography with the spatial constraints of confinement and containment, Horn moves beyond the realm of domestic space, across the globe, to observe, to chart those traces of an inner geography hidden behind the outer busy world of action, change, and ongoing movement. Horn, in her own words stresses that she "uses the outer world to notate the inner." Artistically, she offers a revisionist perspective on Dickinson and her work which interrogates the intensity of her solitariness even as it visually attests to the power of stillness, of contemplation.

Roni Horn's work demands of audiences direct engagement. We must look both backwards to see the impact of Dickinson and forward into the unique artistic representation of that impact. Horn's work calls attention to the primacy of solitude even as it reminds us that the outcome of engaged solitude is an intensification of a sense of togetherness. In the writings of religious mystics globally there is a constant emphasis on the way in which solitude enables solidarity. Theologian William McNamara clarifies this understanding: "Discreet solitude is a creative protest against the euphoric or chaotic togetherness that stamps our way of life in the modern world. But it is also the highest and most apt expression of our solidarity with the whole human race, with the whole of creation. The more solitary we are, the more divinely endowed and psychologically equipped to enter a significantly profound relationship with all levels of life—animal, vegetable, mineral as well as human." Horn seeks to capture that point of entry, to explore the unseen aspects of the contradictions between our desire for connectedness and our longing for spaces that are solitary. Acknowledging both her connection to Dickinson and the way that bond enables a movement out—across boundaries—in *When Dickinson Shut Her Eyes* (1993) she creates a journal-like-narrative recording: "For the time being, Dickinson's here with me, in Iceland. For someone who stayed home she fits in naturally into this distant and necessary place. Her writing is an equivalent of this unique island; Dickinson invented a syntax out of herself, and Iceland did too. Volcanos do. Dickinson stayed home to get at the world. But home is an island like this one. And I come to this island to get at the very center of the world." From this location, at the center, Horn is able to look into the heart of the matter and envision a visual montage of landscapes that allow all of us to come closer to Iceland, to that inner island, without ever having to leave the spaces we call home. This act of remembering, of connecting that which appears to be disparate, separate echoes the thoughts expressed in the last stanza of Dickinson's poem "The Forgotten Grave":

Winds of Summer Fields
Recollect the Way—
Instinct picking up the Key
Dropped by memory—

Horn's work gives us a key—a way to open up the doors of our memory so that a space exists where we travel in and through time.

Her work evokes the poetic lyricism of a past that merges with the present so effortlessly that there is no separation in time. Every moment is a present moment. And it is there in that complete and utter surrender to the present that life is found. It is their devotion to the present moment that is the metaphysical link between Dickinson and Horn. Within

Buddhist teachings, works like the Avatamsaka Sutra remind us that returning to the present is the only way to be in contact with life. Interpreting the sutra Thich Nhat Hanh shares that "time and space are not separate" that "time is made up of space, and space is made up of time." Horn's sculptures urge the onlooker to engage fully with the present. To read the scripts—the messages passed through time—we must be fully aware, mindful in that moment when we confront the art itself. They draw us into the tranquillity of stillness as we observe deeply the image. The joining of physical objects with narrative is a conceptual strategy that calls our attention to dialogical acts of communication that are sacramental.

Indeed, as we look at this work, we bear witness to the communion of souls that takes place between Dickinson and Horn. The reality that this communion remains vibrant and interactive even though these two artists never met in the flesh challenges the conventional notion that death ends communication. The Harlem Boys Choir sings a song with the lyrics "there is no death for an angel." There is definitely an insistence on the primacy of the immortal in Horn's work, substances that will not lose their properties through time but merely undergo various conversions. That hint of immortality, both in the emotional bond she feels with Dickinson and in the work, the traces of love that linger, bind her both to worlds she has never known and those brave new worlds she journeys to. That palimpsest is a mapping of an inner geography that is such a powerful force, it can bring together three women from diverse backgrounds with different histories, who walk different paths and connect us—Dickinson, Horn, Hooks. In our shared immersion in the sacrament of solitude we are renewed, our spirits uplifted. Henri Nouwen writes in his meditation "Out of Solitude" that the experience of solitude enables us to "slowly unmask the illusion of our possessiveness and discover in the center of our own self that we are not what we can conquer, but what is given to us." Horn's work exults in the sharing of these gifts.

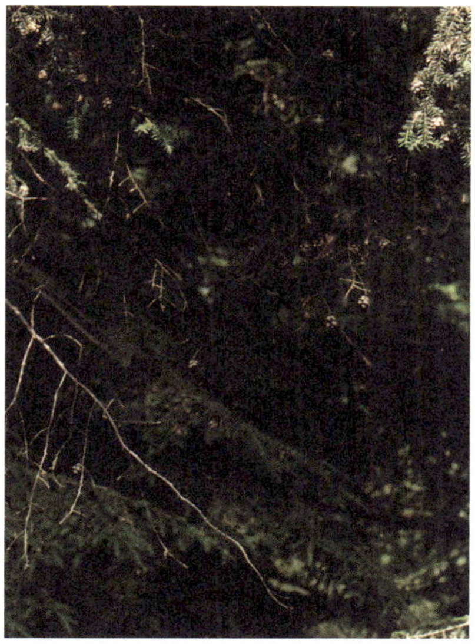

1990: L.A., "The Gold Field"

Felix Gonzalez-Torres

1990: Already ten years into trickle down economics, a rise in cynicism, growing racial and class tension, and the widening gap between the very rich and the rest of us. L.A. before the riots of 1992. A time of defunding vital social programs, the abandonment of the ideals on which our country was supposedly founded. The erasure of history. The Savings and Loan bailout with our tax dollars. "The economic boom" of the Reagan Empire thanks to the tripling of the national deficit. The explosion of the information industry, and at the same time the implosion of meaning. Meaning can only be formulated when we can compare, when we bring information to our daily level, to our "private" sphere. Otherwise information just goes by. Which is precisely what the ideological apparatuses want and need: "You give us 30 minutes and we give you the world." A meaningless one. With internets and highways included. A virtual state of containment. How could the famous "taxpayer" remain idle and voiceless when for every dollar we spent for welfare the government spent six dollars for the Savings and Loan orgy? That was our money that ended up paying for mega takeovers, super mergers, environmentally destructive "developments," and bigger and better (now empty) office spaces. Because it does not mean anything. The American family doesn't know how to get upset, how to understand a bill of 500 billion dollars. But it can get extremely interested in a 10 thousand dollar grant from the NEA. 500 billion is unthinkable. That amount is not personal. On the other hand, 10 thousand dollars is the down payment for a small home, or a trailer. Now that's meaning.

The list of hard data from that fabulous decade was depressing. Especially in the face of public inaction, and the absence of an organized reaction to so many devastating statistics such as the fact that in 1980 the ratio of the U.S. Government budget for housing to military expenditure was 1:5. By 1989 it was 1:31. Since 1980 Federal support for housing assistance has been slashed by more than 80%. According to the Census Bureau, mobile homes were the fastest growing type of dwelling in the 1980s, as the cost of traditional houses soared beyond the reach of many. Nearly 16 million Americans—about one in sixteen—now live in mobile homes. But in a way we built a lot of "safe housing" during that time. According to the September 13, 1992 *New York Times*, the nation's incarcerated population increased by nearly 130% over the decade. We have the highest rate of imprisonment of any industrialized nation. During those same years we witnessed the top one percent of American households grow richer. By 1989 that 1% was worth more than the bottom 90%. In those "Dynasty" years the number of children living in poverty increased by 21%. By 1992, 7% of all infants, and nearly 17% of African-American infants were born underweight—the highest rate since 1978. The state with the highest child poverty rate is Mississippi—home of the American Family Association, one of the most vocal organs of the Right Wing Religious Industry.

According to Jennifer Howse, who led the March of Dimes Birth Defect Foundation in 1992, the proportion of pregnant women without prenatal care was 25%, the highest in 20 years. We now rank 20th among industrialized nations in preventing infant mortality, and when it comes to immunizing infants against polio, we now rank behind 16 other nations, including Mexico. But who cares. We kicked butt in Grenada. We got more stealth bombers, plans for a real star wars system of toys, and a glamorous, anorexic first lady who exhorted all of us to "just say no." To say no to the formulation of meaning, to concentrate instead on a photograph of two men kissing, or of a crucifix. Symbolism sells. History doesn't. We must remember that most of the social security net the revisionists wanted to dismantle then (and now) helped cut the poverty rate almost in half, and poverty amongst the elderly by an even greater degree. And we must remember that the war on poverty in the 1960s and 1970s brought many needy Americans medical care, food stamps, prenatal and infant care, legal services, college tuition and guaranteed student loans which enabled many of us to forge a better life. (I can attest to that.) Those programs, according to the *New York Times* editorial on May 6, 1992, brought the poverty rate down from 19% in 1964 to 11% in 1973.

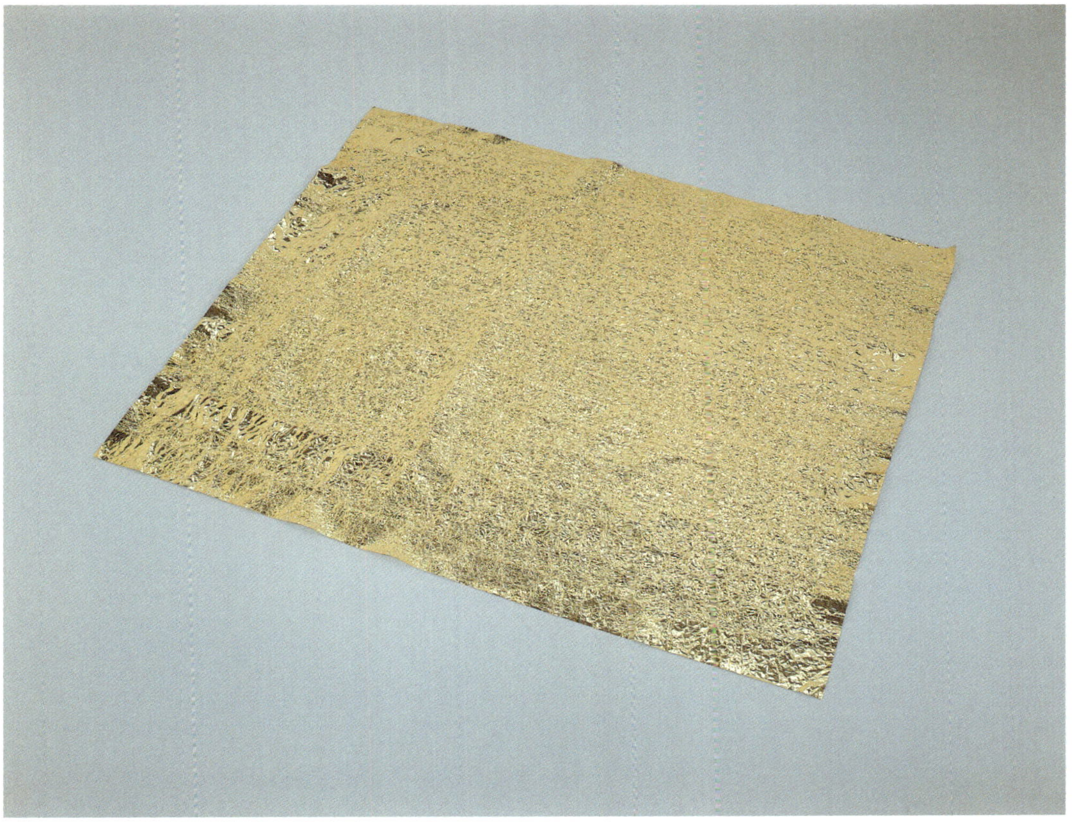

GOLD FIELD, 1982; 4' x 5', pure gold foil

One of the dangers of the new technologies of information is that they do not guarantee an informed or active public. Sound bytes replace arguments. The statistics on the economic decline of the so-called typical normal American family mean very little to them. One of the effects of the division of labor is the misrepresentation of facts, issues, and events as completely isolated, independent of each other. It does not occur to many that bailing out the white collar crimes of the 1980s means less money for hospitals, repairing roads, and school lunches (remember ketchup as a vegetable?).

And it's precisely here where the radical right and their allies in the Religious Industry have been so brilliant in their strategy of deflecting meaning by using charged symbolic images of homosexual acts (among others). Why bother with the destruction of the environment or lack of adequate health care when we have a black and white photo of two men kissing? Now that's real meaning. Unfortunately, we in the cultural left are more than eager to play the role assigned to us. We are invited to participate in a debate that has never really been a debate, but a travesty, a red herring to keep us occupied. We should not reply with the First Amendment and so-called freedom of expression, we should redirect the circus toward our agenda and expose what they really want to avoid mentioning. We should fight hate and the dissemination of ignorance and fear with the effective use of history and fact. Ideology cannot stand it when we make connections.

L.A. 1990. Yes, it was very depressing, and very hard to sustain any sense of hope in such a bleak social landscape. How is one supposed to keep any hope alive, the romantic impetus of wishing for a better place for as many people as possible, the desire for justice, the desire for meaning, and history?

L.A. 1990. Ross and I spent every Saturday afternoon visiting galleries, museums, thrift shops, and going on long, very long drives all around L.A., enjoying the "magic hour" when the light makes everything gold and magical in that city. It was the best and worst of times. Ross was dying right in front of my eyes. Leaving me. It was the first time in my life when I knew for sure where the money for rent was coming from. It was a time of desperation, yet of growth too.

1990, L.A. The Gold Field. How can I deal with the *Gold Field*? I don't quite know. But the *Gold Field* was there. Ross and I entered the Museum of Contemporary Art, and without knowing the work of Roni Horn we were blown away by the heroic, gentle and horizontal presence of this gift. There it was, in a white room, all by itself, it didn't need company, it didn't need anything. Sitting on the floor, ever so lightly. A new landscape, a possible horizon, a place of rest and absolute beauty. Waiting for the right viewer willing and needing to be moved to a place of the imagination. This piece is nothing more than a thin layer of gold. It is everything a good poem by Wallace Stevens is: precise, with no baggage, nothing extra. A poem that feels secure and dares to unravel itself, to become naked, to be enjoyed in a tactile manner, but beyond that, in an intellectual way too. Ross and I were lifted. That gesture was all we needed to rest, to think about the possibility of change. This showed the innate ability of an artist proposing to make this place a better place. How truly revolutionary.

This work was needed. This was an undiscovered ocean for us. It was impossible, yet it was real, we saw this landscape. Like no other landscape. We felt it. We traveled together to countless sunsets. But where did this object come from? Who produced this piece that risked itself by being so fragile, just laying on the floor, no base, no plexiglass box on top of it. How come we didn't know about her work before, how come we missed so much? Roni's work has never been the darling of the establishment. Of course not. Some people dismiss Roni's work as pure formalism, as if such purity were possible after years of knowing that the act of looking at an object, any object, is transfigured by gender, race, socio-economic class, and sexual orientation. We cannot blame them for the emptiness in which they live, for they cannot see the almost perfect emotions and solutions her objects and writings give us. A place to dream, to regain energy, to dare. Ross and I always talked about this work, how much it affected us. After that any sunset became "The Gold Field." Roni had named something that had always been there. Now we saw it through her eyes, her imagination.

In the midst of our private disaster of Ross's imminent death, and the darkness of that particular historical moment, we were given the chance to ponder on the opportunity to regain our breath, and breathe a romantic air only true lovers breathe.

Recently Roni revisited the *Gold Field*. This time it is two sheets. Two, a number of companionship, of doubled pleasure, a pair, a couple, one on top of the other. Mirroring and emanating light. When Roni showed me this new work she said "there is sweat in-between." I knew that.

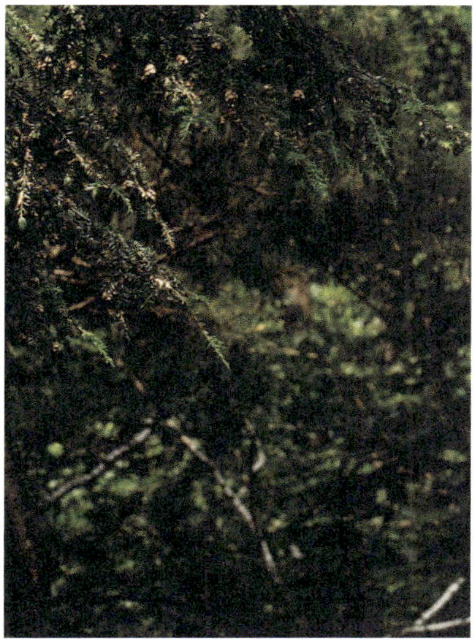

Matrices/Intersections

Judith Hoos Fox

Roni Horn and Judith Hoos Fox have constructed a written dialogue addressing the artist's work inspired by the life and poetry of Emily Dickinson. Fox's questions posed in writing follow. Horn's response forms the text *Among Essential Furnishings*.

The dialogue focuses on a body of work, executed between the years 1992 and 1994, that incorporates Emily Dickinson's writings. Included are two installation works: *How Dickinson Stayed Home* and *Untitled (Gun)*, and two groups of non-installation work—*When Dickinson Shut Her Eyes* and *Keys and Cues*.

1. *How Dickinson Stayed Home,* 1992-93, quotes the phrase: MY BUSINESS IS CIRCUMFERENCE. This text is taken from a Dickinson letter.
2. *Untitled (Gun)*, 1994, quotes the first line of poem No. 754: MY LIFE HAD STOOD—A LOADED GUN—
3. *When Dickinson Shut Her Eyes*, 1993, is a collective title for six individual works, each consisting of the complete quotation of a single Dickinson poem.
4. *Keys and Cues*, 1994, is a collective title for sixteen individual works, each consisting of the quotation of a first line of a Dickinson poem.

In Matrix 1 *the relationship of Horn's work to Emily Dickinson's poetry is explored; in* Matrix 2 *the connections between Horn's Dickinson work and her other work are considered.*

Matrix 1

I

In this body of work your sculptural concerns intersect with Emily Dickinson's poetry. The intersection is, in fact, actual as letter forms comprise one matrix in these bars and blocks of aluminum. Dickinson scholar Susan Howe wrote of the sculptural quality of Dickinson's manuscripts: "In the precinct of Poetry, a word, the space around a word, each letter, every mark, silence, or sound volatizes an inner law of form—moves on a rigorous line" (quoted by Margaret Dickie in "Dickinson in Context," *American Literary History,* vol. 7, no. 2, Summer 1993, p. 325). Nancy Spector wrote of the poetic quality of *Pair Field* and other doubled objects as "analogous to the refrains of a poem" (Nancy Spector, *Roni Horn: Being Double*, Tilburg, The Netherlands: De Pont Foundation for Contemporary Art, 1994, p. 59). That your sculpture would intersect with Dickinson's poetry seems almost preordained. What happens when two distinct voices, expressions in different media, intersect; when text leaves a page and becomes an object; when object contains language itself?

II

Describe your selection process: what criteria led you to choose the particular poems by Emily Dickinson in this body of work? What is the relationship between the poem and the format of the work—the first lines in the *Keys and Cues*; the complete poems, line-by-line, in *When Dickinson Shut Her Eyes*; the stacked blocks in *Untitled (Gun)*; and the single scattered letters in *How Dickinson Stayed Home*?

III

Dickinson writes of holding the firmament within a basket (No. 352), of one bee and a clover forming a prairie (No. 1755). From her cloistered life in Amherst, she precisely expressed universal experience. Does this concept of the part representing the whole hold a place in your thinking? What is the relationship between an excerpted single line and the complete poem; between the single letter and the word it helps to form?

IV

Dickinson wrote her poems on small individual sheets which she tied together into fascicles. The relationship of one poem to another is thus not fixed; context is fluid. Your Dickinson pieces are also presented in segments, letters stacked or scattered, lines leaning. What do these parallels in form suggest? Are meaning and content altered with each installation, as it is in the *Pair Objects*, as it might be in a re-sequencing of Dickinson's poems?

V

For the most part Dickinson wrote her poetry in isolation, without thought of its publication. The irony in the plethora of scholarship surrounding the poet is the fact of making public and accessible the life of one so markedly reclusive. Your Dickinson work seems to confront this conflict directly. The texts are at once evident yet buried, both legible and illegible depending upon the orientation of the leaning bars, the stacked or scattered cubes. Letters, words become an unreadable code, an indiscernible pattern as one views the bars and blocks from various angles. You embrace the simultaneous public and private nature of Dickinson's work and the conflict of their co-existence. What other questions of content does this rendering of text, both eloquent and inscrutable, suggest?

VI

Could the present tense which predominates in both your writing and the poet's indicate a shared commitment to define the nature of experience, the definition of place, the content of existence?—all questions regarding the definition of being here and now.

Matrix 2

VII

This coherent body of work links integrally with concerns you have addressed over the last two decades: the exploration of one kind of experience through another. *Pair Field* explores the experience of memory through physical objects twinned, but placed in separate settings: memory through material. In *Things That Happen Again* four sets of identically twinned objects were located in four varying arrangements, each within its own room or set of rooms. There the objects are the catalyst for the perception of space: space sensed through the object. You have written "(A) place: condensation of acts" (*Roni*

Horn, Munich: Kunstraum, 1983, p. 100) and thus, Iceland has become somewhere known, for those of us who have never been there: place created through word and image. In the Dickinson work you address the abstraction of poetry, of language. Is this metamorphosis akin to other translations, transmutations of sensory experience?

VIII

A consistent investigation in your work has been a search for reconciliation between the interior and the exterior of things, of experience, of objects, a search that raises both formal and philosophical questions. In Iceland, with its chasms, hot springs and raw geography, the interior of the land is revealed as part of its surface. In working with solid materials—copper, aluminum, gold—the surface of your sculptural objects is no different from their interiors; conversely, yet achieving the same end, gold foil is close to being at once surface and interior *(Forms from the Gold Field,* 1980-82). The *Hollowed Masses* and *Bowl of Mass* (1982-83) explicitly address this question. Even in the drawings pigment, image and surface become a unified whole. Physically the Dickinson works, with their letters extending throughout the aluminum bars also contain this integrity between surface and mass. Beyond the physical, what is the meaning of this synthesis between interior and exterior in the Dickinson work?

IX

Through multiplication and repetition you created a seemingly endless vocabulary of forms in *Pair Field* that as a group speak of parity rather than variety. The series of photographs of Margrét, as thought and emotion move through her being, register as a composite of the essential person, rather than of a multifaceted personality. The many strips that comprise the plate drawings unite into a solidified unit. An arrival at the essence of a thing, person or place has been achieved through multiplication and repetition in these and other works. What is the essence you are seeking in Dickinson's poetry?

X

In reference to *Pair Object (I)* you wrote "That the viewer performs the object," (*Roni Horn*, Munich: Kunstraum, 1983, p. 100). The act of perception was one role in which the viewer was directed to engage. How has the role of viewer been amplified, or altered in these works, their content being both literal and literary, formal and philosophical, abstract and specific?

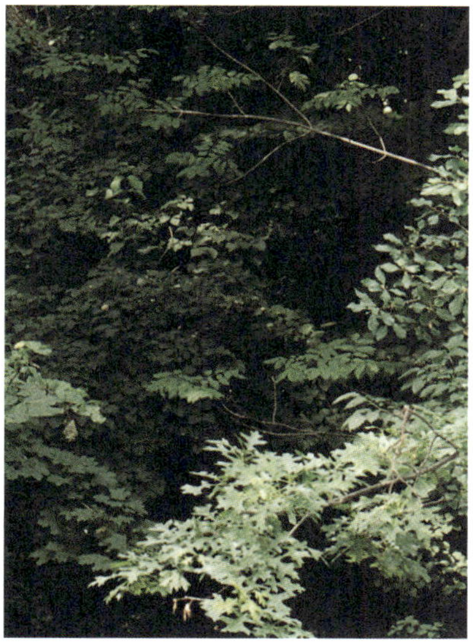

Among Essential Furnishings

Roni Horn

I

I hooked into Emily Dickinson many years ago. Initially it was difficult for me to get past a certain point when reading her. I read her and read her again and it all seemed too simple. Just the same, it was the kind of miss that lingered and teased. Some years later I decided to take a volume of her poetry, the complete works, to Iceland with me. It was the only book I took—I didn't want any options. I spent my evenings in hotels: various rooms in remote and more or less picturesque and barren, landmarkless places. I was alone. It was Dickinson and me. It was the time of year in Iceland when the light of day lingers on into the evening, overcoming the darkness of night. But even in the light of these undarkened nights there was a darkness and solitude that allowed me. It allowed me—just that; and I read Dickinson.

Now as I sit here in New York City confronting the subject of Emily Dickinson I think of the unusual facts of her life and work. Some of those facts impress me deeply, and though I do not know precisely how, they bear directly and indirectly on the extensive group of works I have produced in response to Dickinson's life and poetry.

Dickinson's language grew from the word as written not as spoken. Both her syntax, which is often impossible to predict when reading aloud, and the use of certain words, for example the word 'earths,' in "For Earths, grow thick as / Berries..." (from No. 352) prompt this understanding. —There is a notable visual distance between the handwritten poetry in manuscript form and the typeset material which we read in published form. —Only later in life, when her writing was largely completed, did she fix the relationship among her poems. She sewed them together by hand into fascicles, perhaps more a gesture of binding, than of organizing, perhaps a kind of ritualized private publication. —Dickinson's poetry was written in *private*. By this I don't mean simply that she went off to her bedroom and closed the door and wrote. I mean that no one, not even those who shared her father's home with her throughout her life, *knew the extent of her efforts.* The 1,775 poems she wrote were discovered only after her death.

Only seven poems were published during her lifetime. This silence, willful or not, profoundly influenced her writing. The silence and solitude that pervade her poetry are of a depth and breadth that is new and unfamiliar. Moreover, this unfamiliarity is particular: no amount of familiarity with Dickinson's writings diminishes it. This is one of the qualities that distinguishes her work from all other writing I am familiar with.

The fact that Dickinson did not publish her poetry had a radical though obscure influence on its nature, the direction her writing took, and the realm of heightened sensibility she articulated in the poems. Dickinson's silence is not relative to sound, nor is her solitude relative to the social. Dickinson's silence and solitude were propagated—invented, in the sense that they are so unfamiliar—within the expanse of her relationship to herself. I wonder if you can see her as both the viewer and the view: though the world was vital to her invention, it played a different role in her work than it does for most artists; her view was never mediated by an audience. She only briefly entertained the idea of a larger audience and abandoned it early on. The audience that publication brings, along with its more familiar manner of dialogue, was not a part of her method. Dickinson developed and completed her work within the boundaries of a rare self-sufficiency. In her work the idea of the world exists in the context of this independence. Her self-sufficiency was catalyzed further by this silence/solitude and is perhaps one of the sources for the unique complex knowledge typical of her writing. We know this knowledge did not come from worldly experience. Sixty-nine years after her death her work was made available to the general public in relatively complete form.

II

Here is a fact: there's a lot out there in the world that's not visible either by nature or location. But these non-visible things affect you as deeply as the visible ones. This fact has guided me in developing many of the forms in my work.

To have confluence between exterior and interior, between visible and non-visible, is a form of transparency. This is why the Dickinson works are made the way they are: solid plastic and aluminum, and with an exterior aspect that bespeaks the interior one. I want to emphasize the importance of sensual experience—that people know much more than they think they do. Presence is always a matter of the sensual and it is this, the sensual presence of a thing, that anyone can feel. It is the primary point of contact for an experience. It is direct. You don't need any information or specific knowledge to grasp it. Presence appeals to a physical realm of recognition. This quality of presence requires unity between outside and inside, between visible and non-visible.

In my Dickinson works the text is physically integral to the object. It is not simply painted on. The text is a three-dimensional part of the whole, putting it on a level footing with other physical things. In this manner these works are consistent with my non-text work. Presence and synesthetic experience are qualities I've consistently tried to develop as a means of address to the viewer. Non-sensual experience, which I am perhaps too simply referring to as the conceptual, develops from this.

III

The relationship of my Dickinson work to the poet's original material is a complex one. Personally, I have to consider my extensive quotation of her work in the context of doubling or pairing, which has always been a key theme in my work. This especially concerns the group *When Dickinson Shut Her Eyes,* where I quote whole poems intact. This manner of quotation is a special form of repetition. Formally, it is a manner of appropriation, consumption, mirroring. It is also the transformation of one form into another, a shifting of the means of access. This alteration creates the possibility of new points of view.

In *When Dickinson Shut Her Eyes*—(the title comes from a Dickinson letter, where she wrote, "...to close my eyes is travel..."). I translated the syntax, spacing, and structure of each poem as literally as the text itself. The quatrain form typical of her writing (groups of four lines) dictated the positioning of each line/aluminum bar against the wall; larger spaces among the lines define quatrains, smaller spaces denote lines within a single stanza. The manipulation of the individual lines as they are gathered and arranged into the quatrains reflects the unique syntax of her poems. With a little effort, this construction becomes evident to the viewer.

While these objects speak the poems, they do not represent them. They are them. And I wanted to put this substance, as for example "Earths, grow thick as / Berries...," into the physical world in a manner that will sustain its content, increase its visibility, and allow it a certain directness with the dust and circumstance it so deeply engages.

When Dickinson Shut Her Eyes gives her poems a form that allows them to function in a mundane, easily accessed context. I wanted to take the texts from the book and put them out in the world like any other physical thing with visible attributes and a way of being present. I wanted the possibility that *No. 352* could have a useful place alongside a coat rack or couch. The coat rack holds your coats, the couch offers a place to rest, and *No. 352* offers a certain engagement or occupation of the mind. It now functions in a practical context: it is an essential domestic furnishing. Reading/occupation/engagement is no longer limited to the context of the book. *No. 352* is no longer insulated by other words. It now invites experience parallel to that of furnishings and the practical, everyday world I've mentioned. The change in context alters not the meaning of the poem, but the manner in which it is accessed and therefore used. For me it is significant that *Keys and Cues* and *When Dickinson Shut Her Eyes* are the only objects I have made specifically for domestic space.

Aside from *When Dickinson Shut Her Eyes,* which I regard as a group of discrete objects, there are also two installation works, *How Dickinson Stayed Home* and *Untitled (Gun),* in which I quote phrases, fragments. I am thinking of parts gathered into wholes—the fact of the parts and the manner in which they are gathered and used to extend the whole beyond the original. In these installation works the breaking down and specific rearrangement of text is the foundation of the work—these are complex puns in which the visual and physical play off of the verbal content. In the installation *How Dickinson Stayed Home,* I used a phrase from her letters, MY BUSINESS IS CIRCUMFERENCE. Breaking down this text into its component letters allows me to incorporate the viewer in its syntax. The installation forces the viewer into a physically dynamic reading of the text—a movement that is a metaphor for engagement.

Keys and Cues, like *When Dickinson Shut Her Eyes,* are discrete objects. When I was first becoming familiar with Dickinson's poetry in publications, I found myself reading the "Index of First Lines." Since Dickinson did not title her poems, each was posthumously given a number as a reference point. Each of the 1,775 poems is referred to by number or first line. The Index plays an especially important role in locating a Dickinson poem. In reading it, I found myself thinking of the first lines as entrances, tools, maps, signs, connections. I found myself thinking of alphabets, keys, and elemental things, thinking of things that were self-contained yet only beginnings.

From the 1,775 first lines, I selected lines which are complete statements and detached them from the poem. In these instances I saw the statement as self-contained. I don't think the content of the first line carries the content of the poem it was taken from. The content of a *Key and Cue* has more to do with the particularity that distinguishes Dickinson's poetry as a whole. Each *Key and Cue* carries these qualities and speaks as the part that contains the whole, like a fractal—it can't be broken down into something other than what it is. Each *Key and Cue* is there, as it leans against the wall taking up space, functioning with the pragmatic value of furniture.

The first lines I selected had to have a certain containment (closure) and yet a certain complexity that would hold this completeness open. An experience of a *Key and Cue* is the experience of an entrance; but since every entrance is also a point of departure, the *Key and Cue* prepares the viewer for departure. So while the *Key and Cue* is an object present like any other everyday thing, it is also a cue, a prompt, a signal to something that can only be brought here, wherever the viewer is, by the viewer. These are some of the first lines I used in *Keys and Cues.* They are also statements of fact.

AIR HAS NO RESIDENCE, NO NEIGHBOR.

TO MAKE A PRAIRIE IT TAKES A CLOVER AND ONE BEE

A RAT SURRENDERED HERE

THIS QUIET DUST WAS GENTLEMEN AND LADIES

THE ZEROES—TAUGHT US—PHOSPHORUS—

TWO BUTTERFLIES WENT OUT AT NOON—

REMEMBRANCE HAS A REAR AND FRONT—

Dickinson's facts are shocking. She refers, simply and bluntly, to what's happening out there in the world. Her lines are precise—obscure and unthinkable—but obvious, even inevitable. It's not the facts that are surprising—"Two Butterflies went out at Noon—"; it's that she noted this fact, that this fact was something of value to her.

The reading activity in *Keys and Cues* and where the text takes you is somewhere other than where the physical object is—but they're not far from each other either. These parallel and simultaneous layers in the experience address the viewer separately, appealing to different levels of awareness. They take the viewer to different places. And although these places are not the same, they exist in the same place. For me this is both a necessary paradox and a sensible, even predictable, circumstance.

The present tense is powerful. To tune the entrance of a given experience to an individual's now is startling and acutely focused. That focus on the now has always been important to me, and it's one of the places I found mirrored in Dickinson.

I HEARD A FLY BUZZ—WHEN I DIED—

Awareness is primary in Dickinson—awareness as tangible in the present tense even after it has become past tense. Dickinson's writings give me the sense that she lived her death, like living anything else, noting it very much in the manner she observed butterflies or clover (at least she drew no distinction in magnitude among these things). Knowledge in Dickinson's writings is posterior to experience, and experience is always in the present tense. Anything one can know is.

IV

Some time ago I was in a garden looking out among very old trees when I recognized the need that motivated *When Dickinson Shut Her Eyes.* The experience of those trees made manifest a specific reality. It was pure synthesis, where nothing has a hierarchical relationship to anything else and where everything is somehow equally present, equally affecting.

Dickinson's language scraped the symbolic out of a form that is wholly symbolic and caused it to function in a non-abstract, non-figurative manner. When Dickinson's words cohere into ideas, they are at once and incomprehensibly both metaphysical and physical.

In the reading of Dickinson's work, I am in the world as it is. There is even the sense of circumventing perception, point of view—cutting out everything in-between and every distinction between the viewer and the view. Her writing doesn't take you anywhere. It confirms 'here,' as it is, unimagined, even unseen. All of the paradoxes that constitute her language define it as some other form, one I have no name for. Each thing she names—bees, peace, clover—loses symbolic meaning and takes its place in the actual. It is the absenting of the abstract and the figurative that draws me to quote her.

There is no fantasy in Dickinson's writing, there isn't even the fantastic, even when Dickinson speaks of death—it's eyewitness, it's a death that is recognized in as literal a manner as water or dust. There is this insistent although unstated equality among all things she addresses. From another vantage there is this feeling that the way she lived her life created a fine and acute opportunity. By focusing her life so tightly, she engaged the world through simple recognition. And from this recognition she pared down what was out there to each successive moment of attention. As it turned out, that paring down was really a step deeper into consciousness. She invented a form which is no language at all but a Being Here. She scraped the symbolic from language, impossibility that this is—it is one of her inventions. Each moment of her attention contained a world.

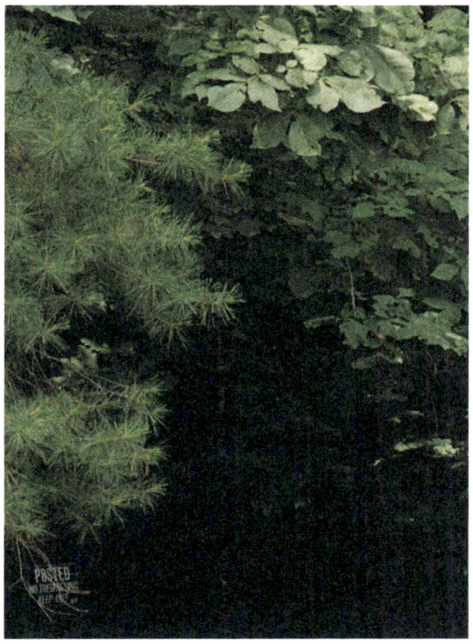

Similar Things in Different Forms

Amada Cruz

For the time being, Dickinson's here with me, in Iceland. For someone who stayed home she fits naturally into this distant and necessary place. Her writing is an equivalent of this unique island; Dickinson invented a syntax out of herself, and Iceland did too. Volcanos do. Dickinson stayed home to get at the world. But home is an island like this one. And I come to this island to get at the very center of the world.[1]

In 1992 Roni Horn began incorporating text from the poetry of Emily Dickinson into a series of sculptures that now number twenty-four works. The affinities between this unlikely pair, the reclusive, nineteenth-century, New Englander Dickinson and the well-traveled, urban, contemporary artist Horn, may at first seem far-fetched. Yet, Horn outlines the similarities in the quotation above. The inventiveness of Dickinson's syntax, her concentrated focus on her immediate world, particularly nature, and the importance of a particular place as a source of inspiration have their parallels in Horn's work.

Emily Dickinson lived a privileged, well-educated, and increasingly solitary life in her home in Amherst, Massachusetts. She began writing verses while very young, sending them to family and friends, and she maintained this prolific correspondence throughout her life. At the time of her death, Dickinson left approximately nine hundred poems in carefully constructed, book-like fascicles held together by thread. Her sister Lavinia discovered this trove while searching through Emily's effects. While her life-long correspondents knew of her poetry, no one imagined the volume of her output.[2] Over the subsequent years, Dickinson's family published her poems and letters, and certain myths about Dickinson developed as a result of the mystery surrounding her life and the romantic nature of some of the poems. She has been portrayed variously as a tortured spinster who retreated from society in reaction to an unrequited love for a married minister, an eccentric proto-feminist who contentedly lived an unconventional life, and a frustrated lesbian in love with her best friend and, later, sister-in-law, Susan. Together, her prose and poetry reveal an inventive and independent poet who found all the stimulus she needed in her library, garden, and home.

Dickinson copied her poems by hand into the fascicles in her unusual style, most often quatrains of short phrases broken up by unclear punctuation and seemingly arbitrary capitalization. The dashes that punctuate her poems are a translation into print of the ambiguous marks she used. It is unclear what her own conversion of these symbols would have been, as she was unsatisfied by the few versions of her poems that she saw published during her life.[3] Dickinson produced different versions of the same verses, and offered variant words within some poems, which she indicated with a cross (+) in her manuscripts.[4] The result of her singular writing technique is a body of work that is fluid and literally open to interpretation. Readers must become active co-authors of her poetry, constructing sentences, deciphering their meanings, and choosing their preferred variants. This method "spotlights the meaning-producing processes of give and take between author and text, text and reader, reader and author, inevitable in reading."[5] The poems then refer back to the way signification is developed, for "on this field of textual play the work becomes a commentary on this process."[6] The poems become about the act of reading them.

A similar integration of form and content exists in the sculptures by Horn that are based on Dickinson's work, the first of which is *How Dickinson Stayed Home* (1992-93). This installation consists of twenty-five aluminum blocks, each five inches cubed, with a letter embedded in each block. These elements are scattered across the floor in an apparently random order. Viewers must traverse this field, mentally connecting letters that assemble a quotation from Dickinson: "My business is circumference." Thus, while constructing the text, viewers inscribe a circumference, also making it their business.

In the next series of Dickinson works, *When Dickinson Shut Her Eyes* (1993), Horn continues her interaction with the viewer. In each of the six works in the series, Horn embeds lines from poems into thin, solid aluminum bars that casually lean against the wall. Each bar consists of one line of text. Some works in the series consist of four lines and just as many bars; others contain seven or eight bars. Horn sets up obstacles to easy reading that recall the disjunctions Dickinson employs in her poetry. Some bars block others from view, and the texts appear correctly on only one of the four sides of each bar. On two of the other sides, viewers see stripes from each letter (that curiously resemble bar codes) and the text printed backwards as a mirror image of the correct version.

In order to make sense of these works, viewers must move around each installation and unravel the text, constructing language and meaning as they interact physically with the works. Although these works are unquestionably sculpture, they are sculptures that must be read, and the text is of primary importance. As such, they are a fusion of sculpture and text, the physical and the conceptual. Horn has balanced these two realms throughout her career, producing installations and books with her writings and photographs of Iceland, and the conceptual concerns of each mode overlap. She has referred to her audience as the "reader/viewer," and has stated, "I'm especially interested in those divergent experiences. That's what I'm after. Similar things in different forms are completely different things."[7]

In some of her books, Horn uses letterpress, a process in which lead type is pressed into paper. She prefers this method because "...there is a greater sense of specificity as to location. Lead type also conveys both textural quality and a very high degree of resolution in each letter... it allows the possibility of typographic drawing."[8] For Horn, the formal considerations of her books, including the physical aspect of the type, are crucial. She treats the texts in her books as she does the letters in her Dickinson works—as a form of drawing that she thoroughly integrates into its support, whether it be paper or aluminum.

From the original handwritten or holographic versions of Dickinson's poems, it is evident that she also sensed the physicality of language. Dickinson's holographs are full of idiosyncratic marks, among them inverted commas, underlines, and the aforementioned crosses. She spaces her words as if to slow down the pace of a poem or accentuate a pause in the flow of text. A particularly noteworthy example is the handwritten version of No. 1210, "The Sea said," in which the poet distributes the letters fairly evenly and far apart across the page and shapes the letters, particularly the "s"es, to look like waves rolling across the manuscript.[9] Readers of Dickinson's holographs thus also become viewers of the surprisingly sensuous script, and the experience of reading this poem, of deciphering language from a field of text, is not dissimilar to the experience of reading/viewing *How Dickinson Stayed Home.*

Keys and Cues (1994) is a series of sixteen sculptures by Horn also inspired by Dickinson's poems. In these works, Horn simplifies the act of reading by presenting only the first line of a poem in each sculpture, which consists of one bar that leans against the wall. Although the text is clearly legible, she obscures its full meaning by omitting the rest of the poem. By isolating one line of a verse, Horn emphasizes the language of that one phrase and leaves the context of the verse up to viewers to imagine. The beauty of Dickinson's language surfaces and, by highlighting one evocative line, an entire range of possibilities emerges.

Horn compounds the physicality of phrases such as "I felt my life with both my hands" (No. 351) with the height and verticality of the bars that compose the *Keys and Cues*, relating them to the human body. She places these works at the edge or circumference of a room, refusing to claim center stage. As with the works in the *When Dickinson Shut Her Eyes* series, viewers encounter the *Keys and Cues* propped up against the wall in an unassuming manner that suggests that the work can be installed in any number of ways. Perhaps the most modest of the Dickinson-based works is *Untitled (Gun)* (1994), a corner piece that, although it stands 11 feet high, appears to be part of the architecture of the room. As in the earlier *How Dickinson Stayed Home*, each block contains a letter, and the parts add up to the first line of a poem. In this case it is from No. 754: "My Life had stood—a Loaded Gun—". As does the sculpture, the loaded gun in the poem stands in the corner until the owner picks it up and carries it away with him. Linda Norden has written that "Horn's sculptures impinge upon their environments not as objects, but as agents or subjects, each modifying the other as they alter our perception of the space they occupy."[10] Rather than being confronted by monumental works, viewers approach Horn's sculptures like landscape elements encountered on a walk—a few works here and there that stop viewers in their tracks long enough to make them notice and contemplate the details.

The landscape of Iceland, which Horn has visited many times since 1975, is an important influence on her work. She has described the rugged, tree-less terrain she encounters on her mostly solitary trips there as somehow necessary to her sense of herself: "Iceland is the place where I have the clearest view of myself and my relationship to the world."[11] In her writings about the island, she characterizes her expeditions as times of heightened awareness. Living simply in a tent, an old lighthouse, or an isolated hotel, Horn drives from one obscure spot to the next, experiencing the peculiar landscape of this young geology with its lava fields, natural pools, glaciers, and desert. "Big enough to get lost on. Small enough to find yourself. That's how I use this island. I come here to place myself in the world. Iceland is a verb and its action is to center."[12]

The last sentence is worthy of Dickinson, who also turned to nature for inspiration. Her poems are full of references to the organic world in earthy details that reveal an acute awareness of her immediate natural surroundings. An example of her practical knowledge (repeated in Horn's *Key and Cue, No. 1755*) is "To make a prairie it takes a clover and one bee." Paula Bennett places Dickinson's preoccupation with nature within the Victorian female tradition of intimacy and identification with the natural world. Through cooking and gardening, women of the time experienced nature in a material and immediate way.[13] Of Dickinson's poetry, Judith Farr says "it attempts to envision the most inchoate, unspeakable, structureless conceptions, such as immortality, in the practical, specific details of everyday." Farr links this vision to the search for the sublime by nineteenth-century American landscape painters such as Thomas Cole and Frederick Edwin Church, noting that Dickinson's chosen pseudonym as a girl was "Cole."[14]

Landscape appears in various guises in Horn's work. Even of the book form she has said, "... reading or looking at a book offers a peculiar analogue both to lived experience and the experience of landscape: sequential relations, mostly cumulative and irreversible, order them."[15] This merging of literature and landscape can be traced to the earliest of her sculptures with text, which she began in 1989. *Thicket No. 1* is the first of these works, an aluminum block that measures 64 x 48 x 2 inches high. Wrapped around its edges is a quotation from Simone Weil: "To see a landscape as when I am not there." A similar work is *Thicket No. 3: Kafka's Palindrome* (1990) with the words: "It would be enough to consider the spot where I am as some other spot." Lying on the floor, these works literally mark a place towards which viewers must walk in order to decode the text. With their insistence upon the viewer's participation, these early sculptures anticipate the Dickinson works.

Horn has stated that reading Dickinson helped clarify the reasons for her periodic trips to Iceland. Horn's search for meaning in nature has been interpreted as a romantic impulse; however, her quest is less a reveling in the wonders of the outdoors than a journey to self-knowledge through direct encounters with a particular place. While Horn travels to Iceland, Dickinson rarely left her house. Yet, the importance of place for both women cannot be overstated. "It's not really about Iceland and it's not really about me. It's about that dialectical interaction with a given place in the world."[16]

Notes

[1]Roni Horn, "When Dickinson Shut Her Eyes," in *To Place: Pooling Waters* (Cologne: Verlag der Buchhandlung Walther König, 1994), p. 71.

[2]Dickinson's oeuvre as presented in the three volumes of *The Poems of Emily Dickinson,* edited by Thomas H. Johnson (Cambridge, Mass.: The Belknap Press of Harvard University Press, 1955), comprises 1,775 poems.

[3]Martha Nell Smith, *Rowing in Eden: Rereading Emily Dickinson* (Austin: University of Texas Press, 1992), pp. 11–14.

[4]Paula Bennett, *Emily Dickinson: Woman Poet* (Iowa City: University of Iowa Press, 1990), p. xiii.

[5]Nell Smith, *Rowing in Eden*, p. 52.

[6]Nell Smith, *Rowing in Eden*, p. 56.

[7]Roni Horn interviewed by Jan Howard, *Inner Geography* (The Baltimore Museum of Art, 1994), p. 29

[8]Roni Horn, interviewed by Jan Howard, *Inner Geography*, p. 28

[9]Nell Smith, *Rowing in Eden*, pp. 83–85.

[10]Linda Norden, "Roni Horn," in *Similia/Dissimilia* (Düsseldorf: Staatliche Kunsthalle and New York: Rizzoli International Publications, 1987), p. 119.

[11]Roni Horn, interviewed by Jan Howard, *Inner Geography*, p. 24

[12]Roni Horn, "Island and Labyrinth," in *Pooling Waters*, p. 23.

[13]See Paula Bennett, "Vinnie's Garden," in *Emily Dickinson: Woman Poet*, pp. 84–115.

[14]Judith Farr, *The Passion of Emily Dickinson* (Cambridge, Mass., and London: Harvard University Press, 1992), p. 248.

[15]Roni Horn, interviewed by Jan Howard, *Inner Geography*, p. 25.

[16]Roni Horn in "Poetry in Place," *Vogue* (September 1993), p. 342.

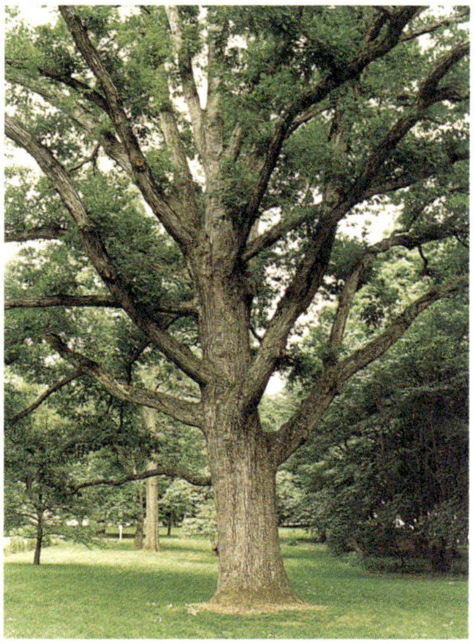

Places between Words

Sarah J. Rogers

The secret of the human condition is that there is no equilibrium between man and the surrounding forces of nature, which infinitely exceed him when in inaction; there is only equilibrium in action by which man recreates his own life through work.

Simone Weil, "The Mysticism of Work," *Gravity and Grace*

Reading Dickinson, I had the experience of remembering things I never experienced.

Roni Horn, *Inner Geography*

What does it mean to take another artist's art as subject and object of your work, especially when the artist in question has achieved considerable renown? For centuries, artists have responded directly and indirectly to the actions and thoughts of other artists, both forebears and contemporaries. Today, however, artists investigating issues of originality, identity, and appropriation tend to create relationships based on mimicry and irony rather than on inspiration. In the case of Roni Horn and Emily Dickinson, where only one artist is living, is this a posthumous collaboration or a strategic appropriation?

Does Horn's use of Dickinson's words form a narrative extension of the poems' meanings or, perhaps, a broader critique of the circumstances of the nineteenth-century American poet, now heralded as one of the geniuses of American literature? None of these summaries truly exhausts the nature of Horn's tie to Dickinson's work. As revealed in Horn's objects, writings, photographs, and books, her relationship to Dickinson transcends such limited partnerships. Rather, what unfolds in a reading of Horn's and Dickinson's art is a series of intricate alliances that run between the bodies of work.

Both are voyagers of sorts, Dickinson through her imagination and words, Horn through the various idioms of her art and her annual sojourns in Iceland. Both are acutely sensitive to the breadth and minutiae of the physical world as evidenced in a Dickinson poem that rhapsodically accounts the approach of night (No. 589) or Horn's gleaming, precisely milled copper *Pair Objects* (1986-88). Both also perceive the syntax and form of language as an aesthetic means to their conceptual desires. But amidst these parallels, questions remain as to why and how Horn has decided to overtly make connections to the life and words of the poet. Pivotal to this response is a strategy that runs through Horn's entire body of work, a strategy both practical and conceptual—that of doubling. It is evident in her use of paired objects and more subtly revealed in the pigment drawings such as the *Distant Doubles* (1988-90). By doubling, Horn explores the relativity and difference of matter—not its uniqueness. In a recent publication, curator and writer Nancy Spector eloquently explains Horn's use of doubling as both empirical and metaphoric, tied to her notion of aesthetic perception, as marking connectors to the real world and the real world behind appearances. She writes, "Horn has created a static form that is emphatically temporal. Its impact is cumulative: the viewer is drawn into a spatial narrative that extends over real time. The visual effect of this experience is stereoscopic: identical images converge in the mind as the eyes encounter duplicate phenomena sequentially."[1]

There are four bodies of works in Horn's Dickinson series. The first, *How Dickinson Stayed Home* (1992-93), is composed of twenty-five five inch aluminum blocks. Embedded in each is a three-dimensional plastic letter. Together the letters spell out the answer to the title's question with a phrase taken from one of Dickinson's letters: "My business is circumference." The second group, entitled *When Dickinson Shut Her Eyes* (1993), comprises six separate pieces, each a grouping of vertical aluminum rods that contain the lines of a Dickinson quatrain. *Keys and Cues* (1994) encompasses sixteen individual works, each a squared aluminum length that contains the first line of a selected Dickinson poem. And the fourth work, *Untitled (Gun)* (1994), stands almost eleven feet high, with twenty-six individual aluminum blocks stacked vertically, each bearing a letter that combine to spell out the first line of one of Dickinson's most renowned poems, "My Life had stood—a Loaded Gun—" (No. 754).

Horn's specific relationship to Emily Dickinson grew over a period of time, most notably during her travels to Iceland when she read Dickinson's numerous letters and poems. Since 1975, Iceland has served as a tactical double to Horn's life in New York. The rugged, geologic landscape still young and changing, the acutely clear atmosphere, the terrain's fissures, pools, and caves have provided a space of physical and intellectual intimacy that then feeds her personal and aesthetic well-being, providing an environment and resource that is a counterpoint to New York City, which is equally necessary to the production of her work. This intimacy is revealed in Horn's connections to literature, whether the writings of Franz Kafka, Simone Weil, Dickinson, or her own written accounts of her solitary wanderings throughout Iceland. Horn's writing style is as concrete and exact as her objects and drawings: restrained, yet embodying a passionate connection to the island of Iceland and her sensorial experiences of that place.

> Dickinson's invention was multiplication, herself, and empirical reach...For the time being, Dickinson's here with me, in Iceland. For someone who stayed home she fits naturally into this distant and necessary place. Her writing is an equivalent of this unique island. Dickinson invented a syntax out of herself, and Iceland did too. Volcanoes do. Dickinson stayed home to get at the world. But home is an island like this one. And I come to this island to get at the very center of the world.[2]

This passage clearly delineates Dickinson's role not merely as inspiration but as a mirror to Horn's own need to travel in order to establish a relationship to a place. Again and again in Horn's art, "place," used both as noun and verb, is the nexus through which empirical knowledge is comprehended.[3] For Dickinson, place is equally pivotal, circumscribed by the contexts of her library, her garden, the view out her bedroom window. Horn uses Iceland as Dickinson used her library, gathering from her travels, her writings and the lens of her camera a means of being centered in the world. These perceptual experiences are then mapped out in her art, in expressive and literal ways. For example, in determining the configuration of an installation, Horn carefully anticipates the viewer's perceptual and physical encounter with these elements: overlaying the overall floor plan with a detailed positioning of the objects. Similarly, Horn's pigment drawings begin with a landscape of forms, rendered as a shape on a two-dimensional surface. These initial drawings are then cut into pieces and collaged back together in new configurations and sequences, as, in a sense, rewritten "texts" of the original drawing (e.g. *Just XVIII*). Horn insists, "I want to see the shape of things from among the things themselves."[4] These strategies overlay and place the individual thing among the whole, forming a landscape of experience that echoes the Dickinson's vision, her ability to perceive the miniature amidst the vast: "To make a prairie it takes a clover and one bee" (No. 1755).

Horn's Dickinson sculptures embody a precision and sensuality found in previous sculptures such as *Gold Mat* (1980-82) or *Asphere III* (1989). In these works the viewer is drawn into an acute perceptual experience by the sheer goldness of the gold or the perfectly imperfect sphere that confounds what we see and what we think we see. Through an absolute insistence on the physicality of the object, Horn's works offer a complexity of meanings to ponder: *Asphere's* perfect, ideal shape that does not conform to the accepted notion of the "ideal." The Dickinson sculptures carry forth this desire to establish an absolute physical presence, using industrial manufacturing processes for their creation. They are also conceptual extensions of the word-based sculptures Horn began in 1989. *Thicket No. 1* (1989) and *Thicket No. 3: Kafka's Palindrome* (1990) are raw and milled aluminum blocks, respectively, plastic letters embedded in the edges spelling out the messages: "TO SEE A LANDSCAPE AS IT IS WHEN I AM NOT THERE" (Simone Weil) and "IT WOULD BE ENOUGH TO CONSIDER THE SPOT WHERE I AM AS SOME OTHER SPOT" (Franz Kafka). The aluminum object, practically and conceptually, is built around a linguistic structure that addresses our perceptual and conceptual experience of what we see in the world. The rectangular block is a space within the space, whose meaning is revealed only as one circulates through the space of and enclosed room. With these word-objects, Horn expresses one of the many connectors to Dickinson's poetry—the power of written language to immortalize gesture and thought. Literary critic Susan Stewart explains,

> Writing gives us a device for inscribing space, for inscribing nature.... Writing serves to caption the world, defining and commenting upon the configurations we choose to textualize. If writing is an

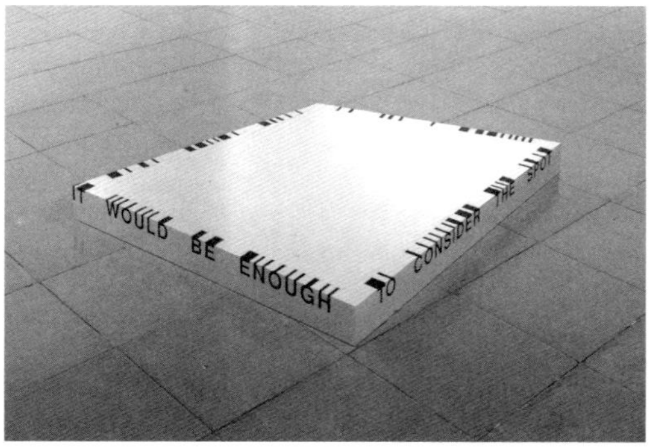

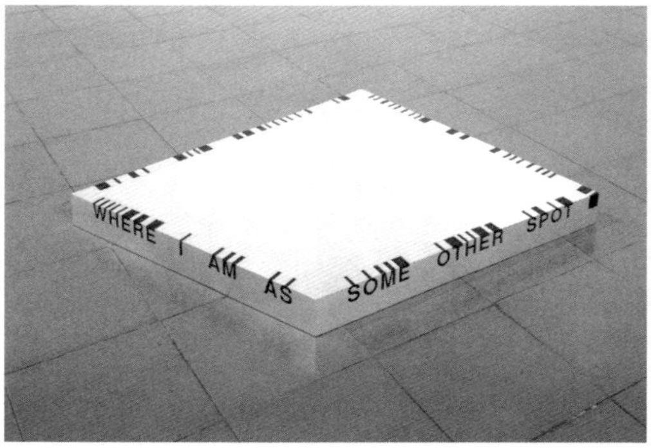

KAFKA'S PALINDROME, 1991
4.25 x 49 x 42 x 40.25 x 36"
solid plastic and aluminum

imitation of speech, it is so as a "script", as a marking of speech taken up through time in varying contexts. The space between letters, the space between words, bears no relation to the stutters and pauses of speech. Writing has none of the hesitations of the body; it has only the hesitations of knowing.[5]

With the Dickinson works, Horn inscribes her sculptural space with language as Dickinson inscribed her personal space through letters and poems. Horn and Dickinson both established an unequivocal relationship to a natural world through the "hesitations of knowing," an essence that then calibrates their equilibrium within the natural world.

A dedication to the provenance of material form and objecthood is also what draws Horn to Dickinson. *When Dickinson Shut Her Eyes, No. 352* is composed of seven aluminum lengths, each embedded with a line of the quatrain, all of which then set together, against a wall. The lengths vary, following the length of the phrase so that the space of the language dictates the space of the object. They lean casually against the wall, a cluster that exists in physical space, just as the Dickinson words , their phrasing, pauses and dashes occupy physical space on the page. The grouping also recalls Dickinson's own bundling of her poems in hand-sewn booklets. "Perhaps I asked too large— / I take—no less than skies—" begins this poem about Dickinson's art and her relationship to nature, two entities inextricably bound. The physicality implied by the construction of words resonates in the physicality of Horn's modest lengths of aluminum. The union of the two brings together the syntax of language and the syntax of form as a strategy for comprehending the world.

Being in the earth: that absolute darkness which is a kind of substance, that exquisite and piercing silence, that loss of relation, that undiluted experience of self which is a self separate from the world.[6]

Sensorial experience is key to the type of knowledge that Horn pursues and that Dickinson sought in her writings. Many Dickinson poems articulate the bodily pleasures of being in the world: seeing the light shift into night, the buzz of a bee, the song of a bird, the scent of a pasture. Each phrase is carefully crafted, no words are unnecessary: "The Day grew small, surrounded tight" (No. 1140) or the sight and sounds of a blue jay "A prompt—executive Bird is the Jay—" (No. 1177). Similarly, Horn's machine milled copper forms, her tactile pigment drawings, her photographs of Iceland's hot springs, all in their own way celebrate the pleasures of hearing, seeing, and touching the world. Horn's series of books *To Place*, most notably the volume *Pooling Waters,* capture the sensual pleasures of being in the landscape. The photographs of geothermal pools, of caves and swimming pools record the effects of water's transmutation into steam or the remarkable clarity of the atmosphere across the vast landscape. The accompanying text recounts the absolute visceral quality of "feeling" Iceland—how Horn felt on her journeys: encountering the wetness, the mud, the instantaneous transitions from hot to cold.

Each how takes you one step deeper, beyond appearance, beyond the simple visibility of things.[7]

As this phrase suggests, the notion of the quest—a journey in the mind or the physical world—is necessary to both artists. Dickinson's remarkable ability to concentrate, removed from the distractions of daily life, enabled her to see and describe a world of exact shapes and senses with great passion ("I felt my life with both my hands," No. 351) and breadth ("An Hour is a Sea," No. 825). For Horn, a journey is inherent in all her art. In *Pair Field* (1991) two identical fields of eighteen unique shapes are positioned in adjacent rooms. The viewer encountering the works is drawn through each space, confronting the pure visibility of things, and then the memory of the first encounter; we are required to experience both form and memory. In Horn's Dickinson works, the poet's words function (symbolically) as that room adjacent to Horn's sculptures. As Horn takes the words and makes them visible in a three-dimensional space, she offers us connectors as we see, read, and feel words, meaning, and experience—so that we remember.

I want to make being here enough.[8]

To be at the nexus of oneself and the world, requires of Horn and Dickinson a solitude that both centers and removes them. But this action is neither mournful nor punitive. For Dickinson, it was a clear, absolute necessity. For Horn, the choice of locating herself at the heart of experience requires a periodic "aloneness" to permit being absolutely present in moment. By their very reductive nature, the *Keys and Cues* objects celebrate this singular, summary of experience. Horn selected the first line of a Dickinson poem for each of the sixteen pieces, again, the text embedded in an aluminum pole that leans against the wall. One reads the phrase as a spine of a book in a library: a good title summarizing the book's contents. Here, Horn found that often the few words in the first line of a poem possessed the power and vision of the entire piece, acting as both the key and the clue to its meaning: "Dust is the only Secret" (No. 153), "I heard a Fly buzz—when I died—" (No. 465), "The Zeroes—taught us—Phosphorus—" (No. 689). The *Keys and Cues* become an absolute, reductive summary of the experience contained in the entire poem.

I want a language without pronouns. I want to come, direct and complete, without pronoun. Yes, I do. I want to come before gender. Yes, yes I do.[9]

This passage by Horn is a poignant summary of the immediacy and passion for a wholeness she seeks to evoke in her objects as well as in her acute sensitivity to language. It also seems to speak for Dickinson in her desire to exist purely through words that mediate the realities of the physical world. In these phrases, Horn longs for freedom from societal constraints that categorize individuals as male or female, heterosexual or homosexual—constraints that thereby enforce assumptions and restrictions. Horn, who very much lives and works in the world, finds momentary relief from these restrictions

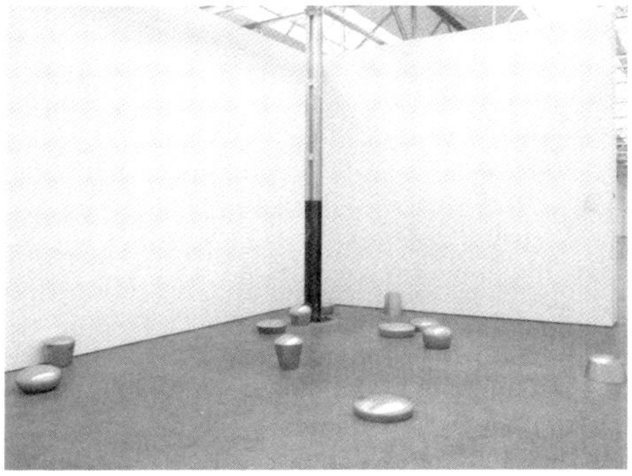

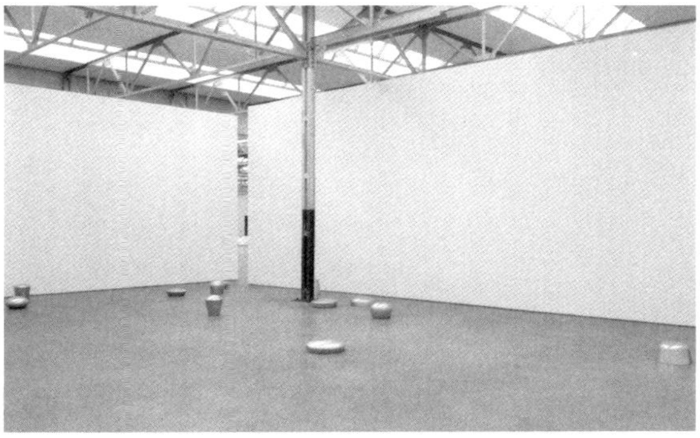

PAIR FIELD—room 1, 1990
solid copper and steel; 18 identical pairs

PAIR FIELD—room 2

when in Iceland, and she carries that freedom back to her work. Dickinson found her own way around these pronouns within the restrictive parameters of nineteenth-century social conventions. She escaped by not leaving her room and by circumnavigating the world through her poetry.[10] Horn, on a deeply personal and artistic level, is connected to such necessity, believing in its unconditional nature. Her Dickinson works, like Dickinson's words, approach an ideal that seeks to unify work, life, and art, and that insists upon the unmediated, visceral reality of the object itself.

Notes

[1]Nancy Spector, "Being Double," *Roni Horn: Being Double* (Tilburg, The Netherlands: De Pont Foundation for Contemporary Art, 1994), p. 58.

[2]Roni Horn, "When Dickinson Shut Her Eyes—I go to Iceland" in *To Place: Pooling Waters*, vol. 2 (Cologne: Verlag der Buchhandlung Walther König), p. 71.

[3]Horn's on-going series of books is entitled *To Place*. To date, the volumes completed are *Bluff Life, Folds, Lava, Pooling Waters,* and *Verne's Journey.*

[4]Roni Horn, "I Can't See the Arctic Circle From Here," in *Pooling Waters,* p. 54.

[5]Susan Stewart, *On Longing: Narratives of the Miniature, the Gigantic, the Souvenir, the Collection* (Durham and London: Duke University Press, 1993), p. 31.

[6]Roni Horn, "Pastoral and Cave," in *Pooling Waters,* p. 62.

[7]Roni Horn, "How—Is Visible Here," in *Pooling Waters,* p. 12.

[8]Roni Horn, "Making Being Here Enough," in *Pooling Waters,* p. 15.

[9]Roni Horn, "Pronouns Detain Me," in *Pooling Waters,* p. 61.

[10]This understanding of Dickinson's reclusive life is fairly recent and in part due to feminist criticism and revisionist readings of her letters and poems, readings that clarify her withdrawal from society as absolutely necessary to her existence, to her poetry. Adrienne Rich's 1975 lecture, later published in *Parnasus: Poetry in Review*, entitled "Vesuvius at Home: The Power of Emily Dickinson" was key to the revised appraisal of Dickinson that sought to understand her not in relationship to a man, but to clearly see her identity as her own, in relationship to herself and to women's traditions and cultures. This essay was important to Horn as she began her own journey through Dickinson's words. Coincidentally, the lecture was given the same year that Horn began her annual sojourns in Iceland.

Afterword: *Presenting a Gift of Presence*

Sherri Geldin

I want to make being here enough.

Roni Horn

It would be enough to consider the spot where I am as some other spot.

Franz Kafka quoted by Roni Horn, *Kafka's Palindrome*

The spot where I am would seem antithetical to Roni Horn, surely unthinkable to Emily Dickinson. Peering from a postmodern turret onto an idyllic nineteenth-century oval grove while leading/coaxing/shaping a center of contemporary art that rigorously seeks its own place in the world. How vast the distance to Amherst or Iceland, even from this rarefied and privileged corner of academe. A spot where daily institutional imperatives so often impede contemplation; a spot where solitude constitutes social offense if not dereliction of professional duty.

Yet being here is enough in some important sense. Enough to give space, time, care, and keen attention to the work of Roni Horn. Work she has summoned from the depths of her singular communion with the poet Emily Dickinson. Work that she's forged with machined precision and unbridled passion Work that she's stripped of all adornment but the essence of seduction itself. Work that strikes with the stark intensity of Dickinson's own.

If, as Horn declares, "Dickinson's facts are shocking," so too are her own. With uncommon ferocity and discipline, Horn pursues her multiple metiers and demands from each its sparest common denominator. That she so relentlessly seeks to find her own center is in itself enough. That she zeroes in on something ineffable and universal is a gift to the rest of us—if we can only find the spot in which to receive it.

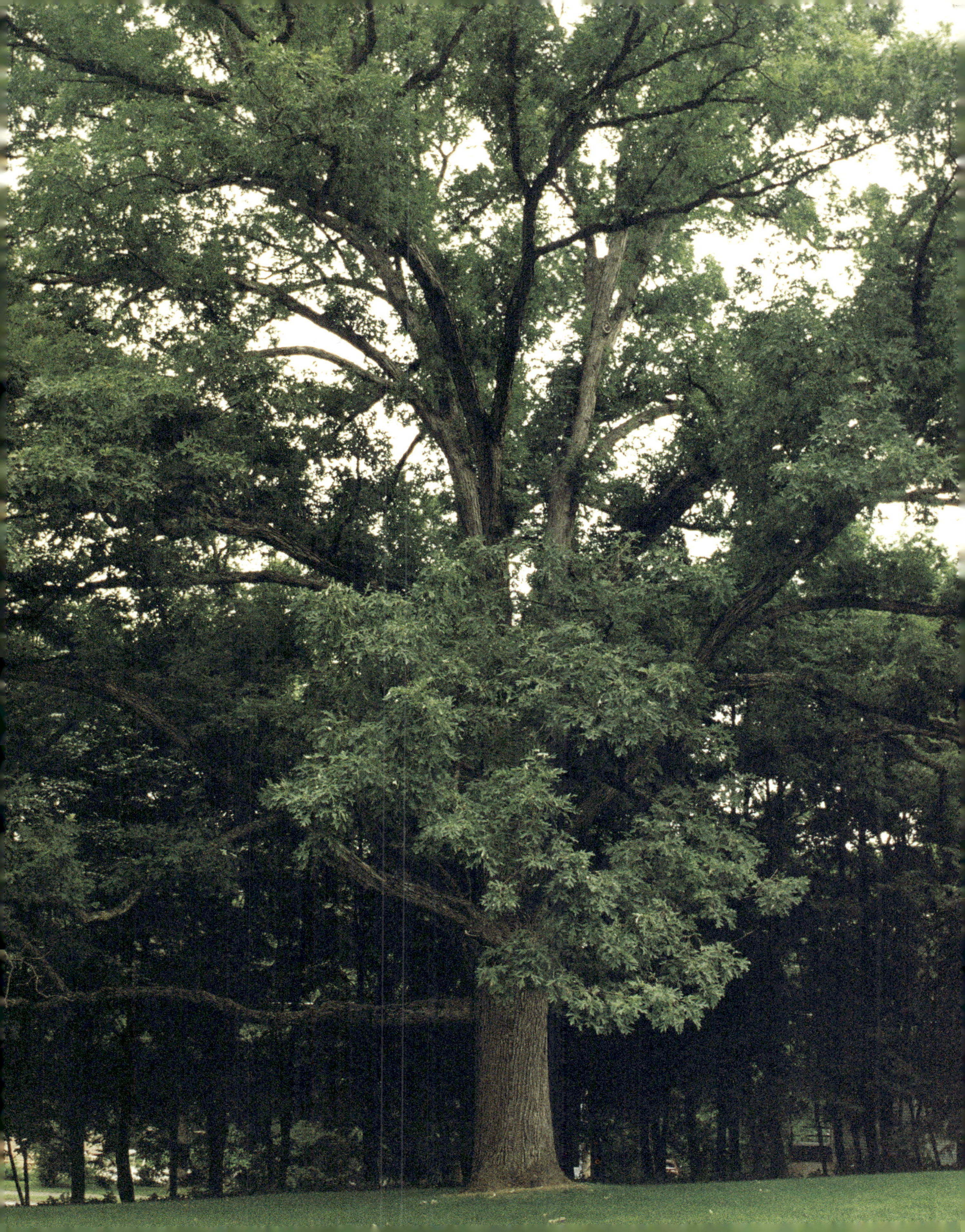

Published on the occasion of the traveling exhibition
EARTHS GROW THICK

Wexner Center for the Arts
The Ohio State University
Columbus, Ohio
February 3–April 21, 1996

Davis Museum and Cultural Center
Wellesley College
Wellesley, Massachusetts
October 22–December 29, 1996

Henry Art Gallery
University of Washington
Seattle, Washington
July 3–September 28, 1997

This exhibition and publication are presented with the generous support of the Lannan Foundation, the Ohio Arts Council, and the Wexner Center Foundation

ORGANIZED BY—Sarah J. Rogers, Director of Exhibitions, Wexner Center for the Arts

ACKNOWLEDGEMENTS—Special thanks to Brenda Richardson for editing *Among Essential Furnishings,* and to Sherri Geldin for her long-standing commitment to my work and the realization of this exhibition.

PHOTO CREDITS—Henk Geraedts, Eindhoven: p. 99; Roni Horn, New York: pp–6-15, 46-55, 57, 65, 71, 77, 85, 93, 103 and cover; Bill Jacobson, New York: pp–21, 22, 29-33, 35, 36, 38, 96; Douglas Parker, Los Angeles: pp–26-27, 67; Nic Tenwiggenhorn, Cologne: pp–17, 24-25.

EDITOR—Ann Bremner, Wexner Center for the Arts
DESIGN—Roni Horn with assistance from Chris Jones

EARTHS GROW THICK—Is published in an edition of 2000 by the Wexner Center for the Arts, Columbus, Ohio. Printed by Hennegan Company, Cincinnati, Ohio. Distributed by D.A.P., New York, New York.

ISBN—1-881390-12-8